T0155912

Springer Graduate Texts in Philosophy

Volume 3

The Springer Graduate Texts in Philosophy offers a series of self-contained textbooks aimed towards the graduate level that covers all areas of philosophy ranging from classical philosophy to contemporary topics in the field. The texts will, in general, include teaching aids (such as exercises and summaries), and covers the range from graduate level introductions to advanced topics in the field. The publications in this series offer volumes with a broad overview of theory in core topics in field and volumes with comprehensive approaches to a single key topic in the field. Thus, the series offers publications for both general introductory courses as well as courses focused on a sub-discipline within philosophy.

The series publishes:

- All of the philosophical traditions
- Includes sourcebooks, lectures notes for advanced level courses, as well as textbooks covering specialized topics
- Interdisciplinary introductions – where philosophy overlaps with other scientific or practical areas.

We aim to make a first decision within 1 month of submission. In the event of a positive first decision, the work will be provisionally contracted. The final decision on publication will depend upon the result of the anonymous peer review of the completed manuscript. We aim to have the complete work peer-reviewed within 3 months of submission. Proposals should include:

- A short synopsis of the work or the introduction chapter
- The proposed Table of Contents
- CV of the lead author(s)
- List of courses for possible course adoption

The series discourages the submission of manuscripts that are below 65,000 words in length.

More information about this series at http://www.springer.com/series/13799

Roman Kossak

Mathematical Logic

On Numbers, Sets, Structures, and Symmetry

Roman Kossak
City University of New York
New York, NY, USA

Springer Graduate Texts in Philosophy
ISBN 978-3-030-07331-2 ISBN 978-3-319-97298-5 (eBook)
https://doi.org/10.1007/978-3-319-97298-5

Softcover re-print of the Hardcover 1st edition 2018

This Springer imprint is published by the registered company Springer Nature Switzerland AG
The registered company address is: Gewerbestrasse 11, 6330 Cham, Switzerland

Preface

Logic and sermons never convince,
The damp of the night drives deeper into my soul.
(Only what proves itself to every man and woman is so,
Only what nobody denies is so.)
Walt Whitman, *Leaves of Grass*

In *Why Is There Philosophy of Mathematics at All?* [8], Ian Hacking writes:

> Yet although most members of our species have some capacity for geometrical and numerical concepts, very few human beings have much capacity for doing or even understanding mathematics. This is often held to be the consequence of bad education, but although education can surely help, there is no evidence that vast disparity of talent, or even interest in, mathematics, is a result of bad pedagogy... A paradox: we are the mathematical animal. A few of us have made astonishing mathematical discoveries, a few more of us can understand them. Mathematical applications have turned out to be a key to unlock and discipline nature to conform to some of our wishes. But the subject repels most human beings.

It all rings true to anyone who has ever taught the subject. When I teach undergraduates, I sometimes say, "Math does not make any sense, right?" to which I hear in unison "Right." My own school experience has not been much different. Even though I was considered "good at math," it only meant that I could follow instructions and do homework to a satisfying result. Only occasionally I'd have moments I was proud of. In my first high school year, I had a mathematics professor about whom legends were told. Everyone knew that in order to survive in his class, one had to *understand*. At any time you could be called to the blackboard to be asked a penetrating question. We witnessed humiliating moments when someone's ignorance was ruthlessly revealed. Once, when the topic was square roots of real numbers, I was called and asked:

– Does every positive number have a square root?
– Yes.
– What is the square root of four?
– Two.
– What is the square root of two?
– The square root of two.

– Good. What is the square root of the square root of two.
– The square root of the square root of two.
– Good. Sit down.

The professor smiled. He liked my answers, and I was happy with them too. It seemed I had understood something and that this understanding was somehow complete. The truth was simple, and once you grasped it, there was nothing more to it, just plain simple truth. Perhaps such rare satisfying moments were the reason I decided to study mathematics.

In my freshmen year at the University of Warsaw, I took Analysis, Abstract Algebra, Topology, Mathematical Logic, and Introduction to Computer Science. I was completely unprepared for the level of abstraction of these courses. I thought that we would just continue with the kind of mathematics we studied in high school. I expected more of the same but more complicated. Perhaps some advanced formulas for solving algebraic and trigonometric equations and more elaborate constructions in plane and three-dimensional geometry. Instead, the course in analysis began with *defining* real numbers. Something we took for granted and we thought we knew well now needed a definition! My answers at the blackboard that I had been so proud of turned out to be rather naive. The truth was not so plain and simple after all. And to make matters worse, the course followed with a proof of existence and uniqueness of the exponential function $f(x) = e^x$, and in the process, we had to learn about complex numbers and the fundamental theorem of algebra. Instead of advanced applications of mathematics we have already learned, we were going back, asking more and more fundamental questions. It was rather unexpected. The algebra and topology courses were even harder. Instead of the already familiar planes, spheres, cones, and cylinders, now we were exposed to general algebraic systems and topological spaces. Instead of concrete objects one could try to visualize, we studied infinite spaces with surprising general properties expressed in terms of algebraic systems that were associated with them. I liked the prospect of learning all that, and in particular the promise that in the end, after all this high-level abstract stuff got sorted out, there would be a return to more down-to-earth applications. But what was even more attractive was the attempt to get to the bottom of things, to understand completely what this elaborate edifice of mathematics was founded upon. I decided to specialize in mathematical logic.

My first encounters with mathematical logic were traumatic. While still in high school, I started reading Andrzej Grzegorczyk's *Outline of Mathematical Logic*,[1] a textbook whose subtitle promised *Fundamental Results and Notions Explained with All Details*. Grzegorczyk was a prominent mathematician who made important contributions to the mathematical theory of computability and later moved to philosophy. Thanks to Google Books, we can now see a list of the words and phrases most frequently used in the book: axiom schema, computable functions, concept, empty domain, existential quantifier, false, finite number, free variable, and many

[1] An English translation is [17].

more. I found this all attractive, but I did not understand any of it, and it was not Grzegorczyk's fault. In the introduction, the author writes:

> Recent years have seen the appearance of many English-language handbooks of logic and numerous monographs on topical discoveries in the foundations of mathematics. These publications on the foundations of mathematics as a whole are rather difficult for the beginners or refer the reader to other handbooks and various piecemeal contributions and also sometimes to largely conceived 'mathematical folklore' of unpublished results. As distinct from these, the present book is as easy as possible systematic exposition of the now classical results in the foundations of mathematics. Hence the book may be useful especially for those readers who want to have all the proofs carried out in full and all the concepts explained in detail. In this sense the book is self-contained. The reader's ability to guess is not assumed, and the author's ambition was to reduce the use of such words as evident and obvious in proofs to a minimum. This is why the book, it is believed, may be helpful in teaching or learning the foundation of mathematics in those situations in which the student cannot refer to a parallel lecture on the subject.

Now that I know what Grzegorczyk is talking about, I tend to agree. When I was reading the book then, I found it almost incomprehensible. It is not badly written, it is just that the material, despite its deceptive simplicity, is hard.

In my freshmen year, Andrzej Zarach, who later became a distinguished set theorist, was finishing his doctoral dissertation and had the rather unusual idea of conducting a seminar for freshmen on set theory and Gödel's axiom of constructibility. This is an advanced topic that requires solid understanding of formal methods that cannot be expected from beginners. Zarach gave us a few lectures on axiomatic set theory, and then each of us was given an assignment for a class presentation. Mine was the Löwenheim-Skolem theorem. The Löwenheim-Skolem theorem is one of the early results in model theory. Model theory is what I list now as my research specialty. For the seminar, my job was to present the proof as given in the then recently published book *Constructible Sets with Applications* [23], by another prominent Polish logician Andrzej Mostowski. The theorem is not difficult to prove. In courses in model theory, a proof is usually given early, as it does not require much preparation. In Mostowski's book, the proof takes about one-third of the page. I was reading it and reading it, and then reading it again, and I did not understand. Not only did I not understand the idea of the proof; as far as I can recall now, I did not understand a single sentence in it. Eventually, I memorized the proof and reproduced it at the seminar in the way that clearly exposed my ignorance. It was a humiliating experience.

I brought up my early learning experiences here for just one reason: I really know what it is not to understand. I am familiar with not understanding. At the same time, I am also familiar with those extremely satisfying moments when one does finally understand. It is a very individual and private process. Sometimes moments of understanding come when small pieces eventually add up to the point when one grasps a general idea. Sometimes, it works the other way around. An understanding of a general concept can come first, and then it sheds bright light on an array of smaller related issues. There are no simple recipes for understanding. In mathematics, sometimes the only good advice is study, study, study..., but this is not what I recommend for reading this book. There are attractive areas of mathematics and its applications that cannot be fully understood without sufficient technical knowledge. It is hard, for example, to understand modern physics without a solid grasp of many areas of mathematical analysis, topology, and algebra. Here I will try to do something different. My goal is to try to explain a certain approach to the theory of mathematical structures. This material is also technical, but its nature is different. There are no prerequisites, other than some genuine curiosity about the subject. No prior mathematical experience is necessary. Somewhat paradoxically, to follow the line of thought, it may be helpful to forget some mathematics one learns in school. Everything will be built up from scratch, but this is not to say that the subject is easy.

Much of the material in this book was developed in conversations with my wife, Wanda, and friends, who are not mathematicians, but were kind and curious enough to listen to my explanations of what I do for a living. I hope it will shed some light on some areas of modern mathematics, but explaining mathematics is not the only goal. I want to present a methodological framework that potentially could be applied outside mathematics, the closest areas I can think of being architecture and visual arts. After all, everything is or has a structure.

I am very grateful to Beth Caspar, Andrew McInerney, Philip Ording, Robert Tragesser, Tony Weaver, Jim Schmerl, and Jan Zwicky who have read preliminary versions of this book and have provided invaluable advice and editorial help.

About the Content

The central topic of this book is first-order logic, the logical formalism that has brought much clarity into the study of classical mathematical number systems and is essential in the modern axiomatic approach to mathematics. There are many books that concentrate on the material leading to Gödel's famous incompleteness theorems, and on results about decidability and undecidability of formal systems. The approach in this book is different. We will see how first-order logic serves as a language in which salient features of classical mathematical structures can be described and how structures can be categorized with respect to their complexity, or lack thereof, that can be measured by the complexity of their first-order descriptions.

All kinds of geometric, combinatorial, and algebraic objects are called structures, but for us the word "structure" will have a strict meaning determined by a formal definition. Part I of the book presents a framework in which such formal definitions can be given. The exposition in this part is written for the reader for whom this material is entirely new. All necessary background is provided, sometimes in a repetitive fashion.

The role of exercises is to give the reader a chance to revisit the main ideas presented in each chapter. Newly learned concepts become meaningful only after we "internalize" them. Only then, can one question their soundness, look for alternatives, and think of examples and situations when they can be applied, and, sometimes more importantly, when they cannot. Internalizing takes time, so one has to be patient. Exercises should help. To the reader who has no prior preparation in abstract mathematics or mathematical logic, the exercises may look intimidating, but they are different, and much easier, than in a mathematics textbook. Most of them only require checking appropriate definitions and facts, and most of them have pointers and hints. The exercises that are marked by asterisks are for more advanced readers.

All instructors will have their own way of introducing the material covered in Part I. A selection from Chapters 1 through 6 can be chosen for individual reading, and exercises can be assigned based on how advanced the students in the class are. My suggestion is to not skip Chapter 2, where the idea of *logical seeing* is first introduced. That term is often used in the second part of the book. I would also recommend not to skip the development of axiomatic set theory, which is discussed in Chapter 6. It is done there rigorously but in a less technical fashion than one usually sees in textbooks on mathematical logic.

Here is a brief overview of the chapters in Part I. All chapters in both parts have more extensive introductions:

- Chapter 1 begins with a detailed discussion of a formalization of the statement "there are infinitely many prime numbers," followed by an introduction of the full syntax of first-order logic and Alfred Tarski's definition of truth.
- Chapter 2 introduces the model-theoretic concept of symmetry (automorphism) using simple finite graphs as examples. The idea of "logical seeing" is discussed.
- Short Chapter 3 is devoted to the elusive concept of natural number.
- In Chapter 4, building upon the structure of the natural numbers, a detailed formal reconstruction of the arithmetic structures of the integers (whole numbers) and the rational numbers (fractions) in terms of first-order logic is given. This chapter is important for further developments.
- Chapter 5 provides motivation for grounding the rest of the discussion in axiomatic set theory. It addresses important questions: What is a real number, and how can a continuous real line be made of points?
- Chapter 6 is a short introduction to the axioms of Zermelo-Fraenkel set theory.

Part II is more advanced. Its aim is to give a gentle introduction to model theory and to explain some classical and some recent results on the classification of first-order structures. A few detailed proofs are included. Undoubtedly, this part will be more challenging for the reader who has no prior knowledge of mathematical logic; nevertheless, it is written with such a reader in mind.

- Chapter 7 formally introduces ordered pairs, Cartesian products, relations, and first-order definability. It concludes with an example of a variety of structures on a one-element domain and an important structure with a two-element domain.
- Chapter 8 is devoted to a detailed discussion of definable elements and, in particular, definability of numbers in the field of real numbers.
- In Chapter 9, types and symmetries are defined for arbitrary structures. The concepts of minimality and order-minimality are illustrated by examples of ordering relations on sets of numbers.

- Chapter 10 introduces the concept of geometry of definable sets motivated by the example of geometry of conic sections in the ordered field of real numbers. The chapter ends with a discussion of Diophantine equations and Hilbert's 10th problem.
- In Chapter 11, it is shown how the fundamental compactness theorem is used to construct elementary extensions of structures.
- Chapter 12 is devoted to elementary extensions admitting symmetries. A proof of minimality of the ordered set of natural numbers is given.
- In Chapter 13, formal arguments are given to show why the fields of real and complex numbers are considered tame and why the field of rational numbers is wild.
- Chapter 14 includes a further discussion of first-order definability and a brief sketch of definability in higher-order logics. Well-orderings and the Mandelbrot set are used as examples.
- Chapter 15 is an extended summary of Part II. The reader who is familiar with first-order logic may want to read this chapter first. The chapter is followed by suggestions for further reading.

In Appendix A, the reader will find complete proofs of irrationality of the square root of two (Tennenbaum's proof), Cantor's theorem on non-denumerability of the set of all real numbers, first-order categoricity of structures with finite domains, existence of proper elementary extensions of structures with infinite domains, and a "nonstandard" proof of the Infinite Ramsey's theorem for partitions of sets of pairs.

Appendix B contains a brief discussion of Hilbert's program for foundations of mathematics.

New York, USA Roman Kossak

Contents

Part I
Logic, Sets, and Numbers

Chapter 1
First-Order Logic

> *However treacherous a ground mathematical logic, strictly interpreted, may be for an amateur, philosophy proper is a subject, on one hand so hopelessly obscure, on the other so astonishingly elementary, that there knowledge hardly counts. If only a question be sufficiently fundamental, the arguments for any answer must be correspondingly crude and simple, and all men may meet to discuss it on more or less equal terms.*
>
> G. H. Hardy *Mathematical Proof* [10].

Abstract This book is about a formal approach to mathematical structures. Formal methods are by their very nature formal. When studying mathematical logic, initially one often has to grit ones teeth and absorb certain preliminary definitions on faith. Concepts are given precise definitions, and their meaning is revealed later after one has a chance to see their utility. We will try to follow a different route. Before all formalities are introduced, in this chapter, we will take a detour to see examples of mathematical statements and some elements of the language that is used to express them.

Keywords Arithmetic · Euclid's theorem · Formalization · Vocabulary of first-order-logic · Boolean connectives · Quantifiers · Truth values · Trivial structures

1.1 What We Talk About When We Talk About Numbers

The natural numbers are 0, 1, 2, 3, ...[1] A natural number is *prime* if it is larger than 1 and is not equal to a product of two smaller natural numbers. For example, 11 and 13 are prime, but 15 is not, because $15 = 3 \cdot 5$. Proposition 20 in Book IX of Euclid's *Elements* states: "Prime numbers are more than any assigned multitude of

[1]According to some conventions, zero is not a natural number. For reasons that will be explained later, we will count zero among the natural numbers.

R. Kossak, *Mathematical Logic*, Springer Graduate Texts in Philosophy 3,
https://doi.org/10.1007/978-3-319-97298-5_1

prime numbers." In other words, there are infinitely many prime numbers. This is the celebrated Euclid's theorem. What is this theorem about? In the broadest sense, it is a statement about the world in which some objects are identified as natural numbers, about a particular property of those numbers—primeness, and about inexhaustibility of the numbers with that property. We understand what the theorem says, because we understand its context. We know what natural numbers are, and what it means that there are infinitely many of them. However, none of it is entirely obvious, and we will take a closer look at both issues later. Concerning the infinitude of primes, it occurred to me once when I was about to show the proof of Euclid's theorem in my class, to ask students what they thought about a simpler theorem: "There are infinitely many natural numbers." It was not a fair question, as it immediately takes us away from the solid ground of mathematics into the murky waters of philosophy. The students were bemused, and I was not surprised.

We will formalize Euclid's theorem in a particular way, and to do this we will have to significantly narrow down its context. In a radical approach, the context will be reduced to a bare minimum. We will be talking about certain domains of objects, and in the case of Euclid's theorem the domain is the set of all natural numbers. Once the domain of discourse is specified, we need to decide what features of its elements we want to consider. In school we first learn how to add and how to multiply natural numbers; and we will follow that path. We will express Euclid's theorem as a statement about addition and multiplication in the domain of natural numbers.

We will talk about addition and multiplication using expressions, called *formulas*, in a very restricted vocabulary. We will use *variables*, two operation symbols: $+$ and $\cdot$, and the symbol $=$ for equality. The variables will be lower case letters x, y, z, ... For example, $x + y = z$ is a formula expressing that the result of adding a number x to a number y is some number z. This expression by itself carries no *truth value*. It can be neither true nor false, since we do not assign any specific values to the variables. Later we will see ways in which we can speak about individual elements of a domain, but for now we will only have the option of *quantifying* over the elements of the domain, and that means stating that either something holds for all elements, or that something holds for some. For example:

$$\text{For all } x \text{ and all } y, x + y = y + x. \tag{1.1}$$

The sentence above expresses that the result does not depend on the order in which the numbers are added. It is an example of a *universal* statement; it declares that something holds for all elements in the domain.

And here is an example of an *existential* statement, it declares that objects with a certain property exist in the domain:

$$\text{There is an } x \text{ such that } x + x = x. \tag{1.2}$$

This statement is also true. There is an element in the domain of natural numbers that has the required property. In this case there is only one such element, zero. But in general, there can be more elements that witness truth of an existential statement.

For example,

$$\text{There is an } x \text{ such that } x \cdot x = x$$

is a true existential statement about the natural numbers, and there are two witnesses to its veracity, zero, and one.

Interesting statements about numbers often involve comparisons of their sizes. To express such statements, we can enlarge our vocabulary by adding a relation symbol, for example $<$, and interpret expressions of the form $x < y$ as "some number x is less than some number y." Here is an example of a true statement about natural numbers in this extended language.

$$\text{For all } x, y, \text{ and } z, \text{ if } x < y, \text{ then } x + z < y + z. \tag{1.3}$$

Notice the grammatical form "if ... then. ..."

The next example is about multiplication. It is an expression without a truth value.

$$1 < x \text{ and for all } y \text{ and } z, \text{ if } x = y \cdot z, \text{ then } x = y \text{ or } x = z. \tag{1.4}$$

In statements (1.1), (1.2), and (1.3), all variables were *quantified* by a prefix, either "for all" or "there exists." In (1.4) the variable x is not quantified, it is left *free*; it does not assume any specific value.

Because of the presence of a free variable, (1.4) does not have a truth value, nevertheless it serves a purpose. It *defines* the property of being a prime number in terms of multiplication and the relation $<$. Let me explain how it works.

Think of a prime number, say 7, as a value of x. If I tell you that $7 = y \cdot z$, for some natural numbers y and z, without telling you what these numbers are, then you know that one of them must be 7 and the other is 1, because one cannot break down seven into a product of smaller numbers. It is true "for all y and z," because for all but a couple of them it is not true that $7 = y \cdot x$, and in such cases it does not matter what the rest of the formula says. We only consider the "then" part if indeed $7 = y \cdot z$. If the value of x is not prime, say 6, then $6 = 2 \cdot 3$, so when you think of y as 2 and z as 3, it is true that $6 = y \cdot z$, but neither y nor z is equal to 6, hence the property described in (1.4) does not hold "for all y and z."

If you are familiar with formal logic, I am explaining too much, but if you are not, it is worthwhile to make sure that you see how the formula (1.4) defines primeness. Chose some other candidates for x and see how it works. Also, notice three new additions to the vocabulary: the symbol 1 for the number one; and two connectives "and" and "or."

With the aid of (1.4) we can now write the full statement of Euclid's theorem: For all w, there is an x such that $w < x$, and for all y and z, if $x = y \cdot z$, then $x = y$ or $x = z$.

What is the difference between the statement above and the original "There are infinitely many prime numbers."? First of all, the new formulation includes the definition of primeness in the statement. Secondly, what is more important, the direct reference to infinity is eliminated. Instead, we just say that for every number w there is a prime number greater than it with such and such properties, so it follows that since there are infinitely many natural numbers, there must be infinitely many prime numbers as well. The most important however is that we managed to express an important fact about numbers with modest means, just variables, the symbols $\cdot$ and $<$, the prefixes "for every" and "there is," and the connectives: "and, " "or", and "if ... then...."

We have made the first step towards formalizing mathematics, and we did this informally. The point was to write a statement representing a meaningful mathematical fact in a language that is as unambiguous as possible. We succeeded, by reducing the vocabulary to a few basic elements. This will guide us in our second step, in which we will formally define a certain formal language and its grammar. We will carefully specify the way in which expressions in this language can be formed. Some of those expressions will be statements that can be assigned truth values—true or false—when interpreted in particular structures. The evaluation of those truth values will also be precisely defined. Some other expressions, those that contain free variables, will serve as definitions of properties of elements in structures, and will play an important role. All those properly formed expressions will be referred to as *formulas*. My dictionary explains that a formula is "a mathematical relationship or rule expressed in symbols." The meaning in this book is different. We will talk about relationships, and we will use symbols, but formulas will always represent statements. For example, the expression $b \cdot b - 4a \cdot c$ is a computational rule written in symbols, but it is not a formula in our sense, since it is not a statement about the numbers a, b, and c. In contrast, $d = b \cdot b - 4a \cdot c$ is a formula. It states that if we multiply b by itself and subtract from it the product of four times a times c, the result is d.

1.1.1 How to Choose a Vocabulary?

In the previous section, we formulated an important fact about numbers—Euclid's theorem—using symbols for multiplication a and the ordering ($<$). This is just one example, but how does it work in general? What properties of numbers do we want to talk about? What basic operations or relations can we choose? The answers are very much driven by applications and particular needs and trends in mathematics. In the case of number theory, the discipline that deals with fundamental properties of natural numbers, it turns out that almost any important result can be formulated in a formal language in which one refers only to addition and multiplication.[2] Number

[2] In our example we also used the ordering relation $<$, but in the domain of the natural numbers, the relation $x < y$ can be defined in terms of addition, since for all natural numbers x and y, x is less than y if and only if there is a natural number z such that z is not 0 and $x + z = y$.

theory may be the most difficult and mysterious branch of mathematics. Proofs of many central results are immensely complex, and they often use mathematical machinery that reaches well beyond the natural numbers. Still, a bit surprisingly, a formal system with a few symbols in its vocabulary suffices to express almost all theorems of number theory. It is similar in other branches of mathematics. The mathematical structures, and the facts about them are complex, but the vocabulary and the grammar of the formal system that we will discuss in this book are much simpler.

The *real numbers* will be defined precisely later. For the moment, you can think of them as all numbers representing geometric distances and their negative opposites. The following statement is written in a rigorous, but informal language of mathematics. It involves the concept of one-to-one correspondence. A one-to-one correspondence between two sets A and B is a matching that to every element of A assigns exactly on element of B in such a way that every element of B has a match.

> Let A be an infinite set of real numbers. Then either there is a one-to-one correspondence between A and the set of all natural numbers, or there is a one-to-one correspondence A and the set of all real numbers.

This is a variant of what is known as the *Continuum Hypothesis*. The Continuum Hypothesis can also be stated in terms of sizes of infinite sets. In the 1870s, Georg Cantor found a way to measure sizes of infinite sets by assigning to them certain infinite objects, which he called *cardinal numbers*. The smallest infinite cardinal number is $\aleph_0$ and it is the size of the set of all natural numbers. It was Cantor's great discovery that the size of the set of all real numbers, denoted by $\mathfrak{c}$, is larger that $\aleph_0$. Another way to state the Continuum Hypothesis is: if A is an infinite set of real numbers, then the cardinality of A is either $\aleph_0$ or $\mathfrak{c}$. The hypothesis was proposed by Georg Cantor in the 1870s, and David Hilbert put it prominently at the top of his list of open problems in mathematics presented to the International Congress of Mathematicians in Paris in 1900. The Continuum Hypothesis is about numbers, but it is not about arithmetic. It is about infinite sets, and about one-to-one correspondences between them. What are those objects, and how can we know anything about them? What is an appropriate language in which facts about infinite objects can be expressed? What principles can be used in proofs? Precisely such questions led David Hilbert to the idea of formalizing and axiomatizing mathematics. There is a short historical note about Hilbert's program for foundations of mathematics in Appendix B.

The Continuum Hypothesis is a statement about sets or real numbers and their correspondences. To express it formally one needs to consider a large domain in which all real numbers, their sets, and matchings between them are elements. Remarkably, it turned out that the vocabulary of a formal system in which one can talk about all those different elements, and much more, can be reduced to logical symbols of the kind we used for the domain of the natural numbers, and just one symbol for the set membership relation $\in$. All that will be discussed in detail in Chap. 6.

So for now we just have two examples of vocabularies, one with symbols + and · for arithmetic, and the other with just one symbol ∈ for set theory. We will see more examples later, and our focus will be on the number structures. In general, for every mathematical structure, and for every collection of mathematical structures of a particular kind, there is a choice of symbols that is sometimes natural and obvious, and sometimes arrived at with a great effort.

A digression: Once formalized, mathematical proofs become strings of symbols that are manipulated according to well-defined syntactic rules. In this form, they themselves become subjects of mathematical inquiry. One can ask whether such and such formal statement can be derived formally from a given set of premises. The whole discipline known as proof theory deals with such questions with remarkable successes. In 1940, Kurt Gödel proved that the Continuum Hypothesis cannot be disproved on the basis of our, suitably formalized and commonly accepted axioms of set theory, and in 1963, Paul Cohen proved that it cannot be proved from those axioms either. This is all remarkable, and was a result of a great effort in foundational studies.

1.2 Symbolic Logic

Mathematical logic is sometimes called *symbolic logic*, since in logical formulas ordinary expressions are replaced with formal symbols. We will introduce those symbols in the next section. Henri Poincaré, the great French mathematician, who was strongly opposed to formal methods in mathematics, wrote in 1908: "It is difficult to admit that the word *if* acquires, when written ⊃, a virtue it did not possess when written *if*."[3] Poincaré was right. Nothing is gained conceptually by just replacing words with symbols, but the introduction of symbols is just a first step. The more important feature is a precise definition of the grammar of the formalized language. We are going to pay close attention to the shape of logical formulas, and the logical symbols will help. It is very much as in algebra: $x + y$, simply means x *plus* y, and $x \cdot y$ means x *times* y, but if you thought that our formalized expression for Euclid's theorem was complicated, think how complicated it would have been if we did not use $+$ and $\cdot$.

There are many advantages of the symbolic notation. It is precise and concise. One not only saves space by using symbols; sometimes symbolic notation allows one to express complex statements that would be hard to understand in the natural language. The most common symbolic system of mathematics is called first-order logic. It will be defined in this section and it will be extensively used in the rest of the book.

The mathematical notation with all its symbols and abbreviations is the language of modern mathematics that has to be learned as any other language, and learning a

[3]In Poincaré's time, ⊃ was used to denote implication.

language takes time. I will try to limit notation to a minimum, but you have to bear with me. The language that you will learn serves communication among mathematicians well, but the fact that so much of mathematics requires it creates problems in writing about mathematics for non-mathematicians. Moreover, mathematicians have created their language in a rather chaotic historical process without particular regard to the needs of beginners. Alexandre Borovik wrote [6]: "Why are we so sure that the *alphabet* of mathematics, as we teach it—all that corpus of terminology, notation, symbolism—is natural?... We have nothing to compare our mathematical language with. How do we know that it is optimal?"

In the previous section we saw how mathematical statements are written using logical connectives and quantifiers. Now we will be writing them using *logical symbols*. Those are $\wedge$, $\vee$, and $\neg$, representing *and*, *or*, and *not*, respectively. The connectives will be used to group together statements about relations, and those statements will be composed of variables and *relation symbols*. The prefixes of the form "for all x ..." and "there is an x such that ..." are the quantifiers, the first is called *universal* and it will be written $\forall x \dots$; the second is called *existential*, and it will be written $\exists x \dots$ (think of ∀ll, and ∃xist).

There are two ways of introducing relation symbols. One could first define an infinite collection of symbols, and then for each structure choose only particular symbols specific to the structure. This would give us a "one language—all structures" model. Alternatively, one can first make a choice of symbols for a particular structure, or a class of structures, and use only those. The latter "one kind of structures—one language" model does not need some of the small technicalities that the former requires, so we will adopt it. Since at first we want to discuss number structures, we choose the following three relation symbols: A for addition, M for multiplication, and L for the "less than" relation. By the standard convention, regardless of the choice of other relation symbols, the equality relation symbol $=$ is also always included in the vocabulary.

For an important technical reason, we will need infinitely many variables. We will index them by natural numbers: x_0, x_1, x_2, x_3, and so on. Each formula will only use finitely many variables, but there is no limit on the number of variables that can be used. This is an important feature of first-order logic so we have to keep all those infinitely many variables in mind, and from time to time there will be a need to refer to all of them. To simplify notation, we will often drop the subscripts, and we will use other letters as well.

We are used to thinking of addition and multiplication as functions, or operations on numbers. Now I will ask you to think of them as relations.

In full generality, the language of first-order logic includes relation symbols and function symbols, but to avoid some technicalities we will not use function symbols. The word "technicalities" is one of those treacherous expressions that often hides some important issues that the author is trying to sweep under the rug, so let me offer an explanation. In mathematics one studies both functions and relations. We use mathematical functions to model processes and operations. Metaphorically speaking, a function "takes" an object as input and "produces" another object as an output. Addition is a two argument function, the input is a pair of numbers, say 1 and 3, and the output is their sum 4. Functions are useful when change is involved; when, for example, some quantity changes as a function of time. Relations are more like databases—they record relationships. Both concepts have their formalizations, and in mathematical practice the distinction between them is not sharp. A relation can evolve in time; a function can be considered as a relation, relating inputs to outputs. The technicalities hinted at above are the rules that must be obeyed when we compose functions, i.e. when we apply one function after another in a specified order. Those rules are not complicated, but at this level of exposition they would require a more careful treatment. We will not do that, and, since every function can be represented as a relation, the price that will be paid will not be great.

Let us see how addition and multiplication can be represented as relations. As we noted earlier, addition of natural numbers is a function. To each pair of natural numbers, the function assigns a value that is their sum. The inputs are pairs of numbers m, n, and the outputs are the sums $m + n$. Let us name this function f. Using function notation, we can write $2+2 = 4$ as $f(2, 2) = 4$, and $100+0 = 100$ as $f(100, 0) = 100$. We are not concerned here with any actual process of adding numbers, we think of f as a device that instantly provides a correct answer in each case. But addition also determines a relationship between ordered triples of numbers as follows. We can say that the numbers k, m, and n are related if and only if[4] $f(k, m) = n$, or, in other words, $k + m = n$. In this sense, the numbers 2, 3, and 5 are related, and so are 100, 0, and 100, but 0, 0, and 1 are not. Notice that we must be careful about the order in which we list the numbers. For example, 2, 2, and 4 are related, but 2, 4, and 2 are not. Addition as a relation carries exactly the same information as the function f does.

To go further, we must now define the rules that generate all formulas of first-order logic. A formula is a formal expression that can be generated (constructed) in a process that starts with basic formulas, according to precise rules. The definition itself is an example of a formal mathematical definition. It is an *inductive definition.* In an inductive definition, one first defines a basic collection of objects, and then describes the rules by which new objects can be constructed from those objects we already have constructed. The definition also declares that only objects obtained this way qualify.

Here is an example of a simple inductive definition. Everyone knows what a finite sequence of 0's and 1's is. It is enough to see an example or two. Here is one: 100011101. Here is another: 1111111. We recognize such sequences when we see them, but notice that this rests on an intuitive understanding of the concept of finite sequence. The inductive definition will not make any explicit references to finiteness, instead the finite character of the concept will be built into the definition. This aspect is not just a philosophical nicety, it has practical consequences. We use inductive definition in a specific way prove results about the defined concepts. The advantage of inductive definitions is that they give an insight into the internal

[4]The phrase "if and only if" is commonly used in mathematics to connect two equivalent statements.

structure of the objects they define. They show us how they are made in a step-by-step process.

Let us now define sequences of 0's and 1's inductively. We begin by saying that 0 and 1 are finite sequences.[5] These are our basic objects. Then comes the inductive rule: if s is a finite sequence then so are $s0$ and $s1$. Finally, we declare that the finite sequences of 0's and 1's are only those objects that are obtained from the basic sequences 0 and 1 by applying the inductive rule (over and over again).

Now we go back to formulas. Recall that we chose A, M, and L for relation symbols. A and M are *ternary*—they bind three variables, and L is *binary*—it binds two variables. "Binding" is a technical term, and you should not put more meaning to it beyond what is written in the following definition of basic formulas, which we will call *atomic*. Atomic formulas are all expressions of the form $A(x_i, x_j, x_k)$, $M(x_i, x_j, x_k)$, $L(x_i, x_j)$, and $x_i = x_j$, where i, j, and k are arbitrary natural numbers. For example, $A(x_0, x_1, x_2)$, $M(x_5, x_3, x_1)$, $M(x_1, x_0, x_0)$, $L(x_1, x_2)$ are atomic formulas. In other words, an atomic formula is a relation symbol followed by a list of three arbitrary variables, in the case of A and M, or two arbitrary variables, in the case of L. Notice that since there are infinitely many variables, there are also infinitely many atomic formulas.

When we discuss particular examples of formulas, for greater readability we will usually drop the subscripts and use other letters for variables, so expressions such as $A(x, y, z)$ or $L(y, x)$ (not a typo) will also be considered atomic formulas, although in the strict sense, according to the definition they are not.

Along with formal expressions, defined according to strict rules, we will also use other common mathematical expressions and those will also often use symbols. The two different kinds of symbols should not be confused. We will need names for many different objects, including formulas and sentences of first-order logic. We will see this in the definition below. The Greek characters φ (phi), ψ (psi), and other, are used as names for formulas of the language we define. They are not a part of the formalism.

Definition 1.1

1. Every atomic formula is a formula.
2. If φ and ψ are formulas, then so are $(\varphi) \wedge (\psi)$, $(\varphi) \vee (\psi)$, and $\neg(\varphi)$.
3. If φ is a formula, then, for each n, $\exists x_n(\varphi)$ and $\forall x_n(\varphi)$, are formulas.
4. There are no other formulas.

Notice the use of parentheses. They play an important role. They guarantee that every first-order formula can be read in only one way.[6]

---

[5]We could actually start one level lower. We could say that the empty sequence, with no symbols at all, is a finite sequence of 0's and 1's.

[6]This unique readability of first-order formulas is not an obvious fact and requires a proof, which is not difficult, but we will not present it here.

Let us see how Definition 1.1 works. Per clause (1.1), the expressions $x = y$ and $A(x, z, y)$ (no typos here) are formulas, because they are atomic formulas.[7] Per rule (1.2), $\neg(x = y)$ and $\exists z(A(x, y, z))$ are formulas. By applying (1.2), we see that

$$(\neg(x = y)) \wedge (\exists z(A(x, y, z)))$$

is a formula as well. Let us call this formula $\varphi(x, y)$.[8] The only displayed variables are x and y, because they are free in $\varphi(x, y)$. The third variable z is bound by the existential quantifier $\exists z$. Think of A as the addition relation of the natural numbers. For what values of x and y does $\varphi(x, y)$ become a true statements. First of all they must be different, as declared by the first component of $\varphi(x, y)$, but also x must be less than y, because only then there is a natural number z such that $x + z = y$.

An important caveat. According to the rules, we are free to choose any variables we like to form atomic formulas, so for example $A(x, z, z)$, and $L(x, x)$ are well-formed formulas. If the same free variable is used in different places in a formula, when we interpret the formula in a structure, that variable will always be evaluated by the same element, but this does not mean that if the variables are different that they represent different objects. We are free to evaluate any free variable by any object, in particular we can use the same object for different variables.

Let us recapitulate. The list of symbols of the first-order logic is: $\wedge$, $\vee$, $\neg$, $\exists$, $\forall$, and $=$, and then for each particular structure, in addition, it includes a collection of relation symbols. In our case, we chose A, M, and L. Each relation symbol has a prescribed *arity* which is given in the definition of the atomic formulas. The symbols A and M are of arity three, and L is of arity two. This means that, for example, $A(x_0, x_1, x_2)$ is a well-formed atomic formula, but $A(x_0, x_1)$ and $L(x_0, x_1, x_2)$ are not, because the number of variables does not match the arity of the symbol.

The attentive reader will ask: But what about all those parentheses and commas? Yes, they are also formal symbols of our language, and their use is entirely determined by Definition 1.1. The rules for commas are hidden in clause (1.1). One could do without them. For example $Ax_0x_1x_2$ also represents a uniquely recognizable string of symbols, and this is all we want, but for greater readability, and to avoid additional conventions that we would have to introduce in the absence of parentheses, we use parentheses and commas. Often we will also use "[," and "]," and sometimes "{" and "}." They all have the same status as "(," and ")."

[7]This is an example of mathematical pedantry. Of course, you would say, they are formulas. They are even atomic formulas! But when we defined atomic formulas, we defined a special kind of expression, and called expressions of this kind "atomic formulas." When we did that, the formal concept of formula had not been defined yet. To know what a formula of first-order logic is one has to wait for a formal definition of the kind we gave here. To avoid this whole discussion we could have called atomic formulas atoms. If we did that, then clause (1.1) of the definition above would say "Every atom is a formula," but since the term "atomic formula" is commonly used, we did not have that choice.

[8]This is another example of an informal abbreviation.

Each application of rules (1.2) and (1.3) introduces a new layer of parentheses. If we continue this way, formulas quickly become unreadable, but this formalism is not designed for the human eye. We sacrifice easy reading, but we gain much in return. One bonus is that it is now easy to check and correct grammar. The only grammatical rules are those in Definition 1.1. In particular, in every formula the number of left parentheses must be equal to the number of right parentheses. If it is not, the sequence of symbols is not properly formed and it is not a formula.

The most important aspect of Definition 1.1 is that it shows how all formulas are generated in a step-by-step process in which more and more complex formulas are generated. This is a crucial feature, that opens the door to investigations of formal languages by mathematical means. Another essential feature is that the set of all formulas is generated without any regard to what those formulas may express. In fact, most formulas do not express anything interesting at all. For example

$$((x = x) \vee (x = x) \wedge (x = x))$$

is a proper, but uninteresting formula, and so is $\exists x L(y, z)$.

What is the point of allowing meaningless formulas? What we are after are formulas and sentences that express salient properties of structures and their elements, but we would be at a loss trying to give a mathematical definition of a meaningful formula. It is much easier to accept them all, whatever they may be expressing. There is something profound in treating all formulas this way. Meaningfulness is a vague concept. A sentence of no interest today, may turn out to be most important tomorrow, so it would make no sense to eliminate any of them in advance, but this is not the main point. Most mechanically formed formulas are not only uninteresting, they actually make no sense at all. Still we want to keep them in, because it is the price to pay for the clarity of the definition. Moreover, there are also some unexpected technical applications. If φ is a formula, then, according to rule (1.2), so are $(\varphi) \wedge (\varphi)$ and $(\varphi) \wedge ((\varphi) \wedge (\varphi))$, and $(\varphi) \wedge ((\varphi) \wedge ((\varphi) \wedge (\varphi)))$, and so on. Nothing new is expressed, but there are some important results in mathematical logic that depend in an essential way on existence of such statements.

The definition of the syntax of first-order logic is completed. Now it is time to define the *semantics*, i.e. the procedure that gives meaning and truth values to formulas when interpreted in a structure. We already did that informally, when we talked about Euclid's theorem and interpretations of formulas in the natural numbers. Full definition of semantics for first-order logic is based on Alfred Tarski's famous *definition of truth* from 1933 [34]. It is formulated in a set-theoretic setting that we will discuss later. For now, we will show how it all works using examples. For a full formal definition consult any textbook on mathematical logic. A good source online is the Stanford Encyclopedia of Philosophy [13].

We will interpret formulas in the domain of the natural numbers. To begin with, for any three numbers m, n, and k, we need to assign truth values (true or false) to all atomic formulas $A(x, y, z)$ and $M(x, y, z)$, $L(x, y)$, and $x = y$, when x, y, and z are interpreted as m, n, and k respectively. We declare $A(x, y, z)$ to be true if and only if $m + n = k$, $M(x, y, z)$ to be true, if and only if $m \cdot n = k$, $L(x, y)$ to be true if and only if m is less than n, and finally $x = y$ to be true if and only if m equals n. We are exceedingly pedantic here, and for a good reason. We just described the definition of truth for the atomic formulas.

What makes notation complicated in the explanations above is the reference to evaluation of the variables. To simplify matters, one is tempted to assign truth values directly to expressions such as $A(m, n, k)$. There is a problem with that. The expression $A(x, y, z)$ is a formula. It is just a string of symbols of the language of first-order logic. The expression $A(m, n, k)$ is not a formula. The letters m, n, and k, as used here are informal names for numbers. No rule in Definition 1.1 allows inserting names of objects into formulas. In the expression $A(m, n, k)$ two worlds are mixed. The relation symbol A, the parentheses and commas, come from the world of syntax; m, n, and k are not symbols of the formal language, they are informal names of elements of the domain of the structure.

What is the difference between the statement "$A(x, y, z)$ is true, when x, y and z are interpreted as m, n, and k" and the statement "$m + n = k$"? The former states that a certain *truth value* is assigned to a certain formula under certain conditions. The latter is a statement about the state of affairs in a certain structure. While the definition is telling us under what conditions certain statements are true, it has nothing to do with whether we can actually check if those conditions are satisfied. In the case of checking whether m plus n equals k, think of numbers so incredibly large that there is not enough space to write them down. We are not talking of any practical aspects of computation here. Still, it makes sense to define the truth values of interpretations of formulas this way. The definition is precise, and it is exactly this definition that makes a bridge between the syntax and the world of mathematical objects in which it is interpreted.

Once the definition of truth values for atomic formulas is established, truth values for more complex formulas are determined in a way parallel to the rules for generating formulas in Definition 1.1. For example, $\neg(\varphi)$ is true if and only if φ is false; $(\varphi) \wedge (\psi)$ is true, if and only if both φ and ψ are true; and $\exists x(\varphi)$ is true if and only if there is an evaluation of the variable x under which φ becomes true. Here we take advantage of the inductive form of Definition 1.1. In the same way in which the more complex formulas are inductively built from simpler ones, the truth values of more complex formulas are inductively determined by the truth values assigned to their simpler components, with the atomic case as the base. As was mentioned earlier, the full formal definition of this process is somewhat technical, and we will omit it.

In our discussion of Euclid's theorem, we included "if ... then ..." among the logic connectives. Conditional statements of the form "if φ then ψ" abbreviated by $(\varphi) \implies (\psi)$, are essential in mathematics, but Definition 1.1 has no provision for them. One could add another clause there explaining how $(\varphi) \implies (\psi)$ is to be interpreted, but this is not necessary. In classical logic, the formula $(\varphi) \implies (\psi)$ is defined as an abbreviation of $\neg(\varphi) \vee (\psi)$, hence its interpretation is already covered by the Definition 1.1. Let us see it on an example we already discussed.

The formula (1.4) defining prime numbers in the previous section included the following conditional statement:

$$\text{For all } y \text{ and all } z, \text{ if } x = y \cdot z, \text{ then } x = y \text{ or } x = z.$$

Its symbolic version is

$$\forall y \forall z[(M(y, z, x)) \Longrightarrow [(y = x) \vee (z = x)]], \tag{1.4'}$$

which in turn is equivalent to

$$\forall y \forall z[\neg(M(y, z, x)) \vee [(y = x) \vee (z = x)]]. \tag{1.4''}$$

Convince yourself (1.4″) is true only if x is interpreted as either 1 or a prime number.

Another common logical connective is "if and only if." The symbol of it is $\Longleftrightarrow$, and $(\varphi) \Longleftrightarrow (\psi)$ is defined as an abbreviation for $((\varphi) \Longrightarrow (\psi)) \wedge ((\psi) \Longrightarrow (\varphi))$. For an example, see Exercise 1.5.

A first-order property is a property that can be expressed in first-order logic, which means it can be defined by a formula in the formalism we just described. This whole book is about mathematical structures and their first-order properties. Not all properties are first-order. For example, here is a property of natural numbers that is not defined in a first-order way

$$\text{Every set of natural numbers has a least element.} \tag{1.5}$$

In (1.5) we quantify over sets of numbers, and that makes this statement *second-order*. In first-order logic we can only quantify over individual elements of domains, but we cannot quantify over sets of elements. Quantification over sets is allowed in second-order logic with its special syntax and semantics. There is a third-order logic that allows quantification over sets of sets of elements. There are higher-order logics, each with stronger expressive powers. There is more about this in Chap. 14.

We will stick to first-order logic for two reasons. One is that even with its restrictions, first-order logic is a strong enough formal framework for a substantive analysis of mathematical structures in general, but there is also another appealing reason. First-order logic is based on rudimentary principles. It only uses simple connectives "and," "or," and "not," and the quantification only allows us to ask whether some property holds for all elements in a domain ($\forall$), or if there is an element in a domain with a given property ($\exists$). In other words, it is a formalization of the most basic elements of logic, and one could argue that it also captures some basic features of perception. Let's think of collections of elements and some of their properties that can be visually recognized. If I see a set of elements having a property $\varphi(x)$, I also see its complement consisting of the elements that do not have that property, which is the same as having the property $\neg\varphi(x)$. If some elements have a property $\varphi(x)$, and some have another property $\psi(x)$, then I can see the collection of elements with both properties, i.e. the set of defined by property $\varphi(x) \wedge \psi(x)$.

Similarly I can see the elements having one property or the other: $\varphi(x) \vee \psi(x)$. If all elements have a property $\varphi(x)$, I see that $\forall x \varphi(x)$. To see that it is not the case, it is enough to notice one element that does not have the property, so it is enough to see that $\exists x \neg\varphi(x)$. The first-order approach provides a basic framework for what I will call *logical visibility*. Equipped with this framework, we will try to find out what can, and what cannot be seen in structures through the eyes of logic.

Here is a rough outline of what we will do next. To define a structure, we start with a collection of individual objects sharing certain features. In each structure, the objects in the collection are related to one another in various ways. We will give those relations names, and then we will try to see what properties of the structure and its individual elements are first-order. An analysis of the complexity of formulas and sentences of first-order logic will allow us to apply geometric intuitions, and to see geometric patterns in the structure. In this sense, the formalism will allow us to go back to more natural, unformalized ways of thinking about the structure, and to "logically see" some of its features, that otherwise might have stayed invisible. This works well in mathematics and we will examine some examples.

1.2.1 Trivial Structures

The simplest structures are domains with no relations on them. Think of a domain with five objects. If those objects are not related to one another in any way, this is an example of a *trivial* structure. What can be said about it? Not much more than what we have said already, but it is good to keep trivial structures in mind for further discussion. They are a good source of examples and counterexamples. Due to our convention, the equality relation is always among the relation symbols for any structure. Hence, even though a trivial structure has no relations of its own, it still has the equality relation, and it allows us to express specific facts about it in the first-order way. Consider the sentence:

$$\exists x \, \exists y \, [\neg(x = y) \wedge \forall z \, (z = x) \vee (z = y)].$$

It says that there are two distinct elements and any element in the structure must be one of them. In other words, it expresses that the structure has exactly two elements. In a similar way, for any number n, one can write a first-order sentence expressing that the structure has exactly n elements.

It is an interesting fact, that follows from the compactness theorem for first-order logic, that while for each number n, having exactly n elements is a first-order property, having a finite number of elements is not. The compactness theorem and an argument showing why finiteness is not a first-order property are presented in Chap. 11.

Exercises

Exercises marked by the asterisk are more advanced.

Exercise 1.1 *Write first-order sentences expressing the following:*

1. *There are at least three elements.*
2. *There are at most five elements.*
3. *There are either three, four, or five elements.*

Exercise 1.2 *Write the Euclid's theorem as a first-order sentence using the ternary relation symbols A and M and the binary symbol L.*

Exercise 1.3 *Twin primes are prime numbers that differ by* 2*. For example,* 3 *and* 5 *are twin, and so are* 11 *and* 13*. The Twin Primes conjecture says that there are infinitely twin prime numbers. We do not know if the conjecture is true, although there has been recent progress in number theory suggesting that it may be. Express the Twin Primes conjecture by a first-order sentence using the ternary relation symbols A and M and the binary symbol L.*

Exercise 1.4 *The Goldbach conjecture says that every even number greater than two is a sum of two prime numbers. For example:* $4 = 2 + 2$, $6 = 3 + 3$, $8 = 3 + 5$, $100 = 3 + 97$*. Express the Goldbach conjecture by a first-order sentence using the ternary relation symbols A and M, and the binary symbol L.*

Exercise 1.5 * *Later in the book, instead of fully formal expressions, we will use more readable notation. For example, instead of* $A(x, y, z)$ *we will simply write* $x + y = z$*. We will also use abbreviations. In the example below,* $P(x)$ *stands for a first-order sentence expressing that* x *is prime. Fermat's theorem says that a prime number* p *is a sum of two squares if and only if* $p = 4m + 1$*, for some natural number* m*. To express the theorem in a more formal way, one can write*

$$\forall x[P(x) \implies (\exists y \exists z\ (x = y^2 + z^2) \iff \exists t\ (x = 4t + 1)].$$

Try to write this sentence using the relation symbols A, M, and L, as in the previous exercises. It will be long.

Chapter 2
Logical Seeing

Abstract This chapter serves as an interlude. Our goal in the following chapters is to present a formalized approach to numbers, and then we will look at the number systems again to see how tools of logic are used to uncover their essential features. We will be inspecting the *structure* of the number systems with our logic glasses on, but we need to get used to wearing those glasses. In this chapter we will take a look at some simple finite structures—finite graphs—and we will examine them from the logical perspective. In other words, later, logic will help us to see structures; now, some simple structures will help us to see logic. An important concept of symmetry of a graph is introduced in Definition 2.1 followed by equally important Theorem 2.1. Both, the definition and the theorem, will be generalized later to arbitrary mathematical structures.

Keywords Finite graphs · Graph symmetry · Fixed points · Types of elements · Addition and multiplication as relations

2.1 Finite Graphs

A graph is a collection points, called vertices, some of which are connected by edges. Graphs are mathematical structures. The collection of vertices is the domain of the structure, and the vertices are related by the *edge relation*. The edge relation is *symmetric*: An edge between vertices v and w is also an edge (the same edge) between w and v. Loops are allowed. A *loop* is an edge between a vertex and itself. In graph theory one encounters different types of graphs. There are directed graphs in which the edge relation is not symmetric, and there are graphs in which multiple edges between vertices are allowed. To keep the matters as simple as possible, we will not discuss directed graphs, and we will not allow multiple edges.

The first-order language for graphs has one binary relation symbol. We will use E. Thus, the atomic formulas (see Definition 1.1) are all formulas of the form $E(x_i, x_j)$, where x_i and x_j are arbitrary variables. If v and w are vertices in a graph, and x_i is interpreted by v, and x_j by w, then $E(x_i, x_j)$ is true under this evaluation if there is an edge between v and w, otherwise it is false.

R. Kossak, *Mathematical Logic*, Springer Graduate Texts in Philosophy 3,
https://doi.org/10.1007/978-3-319-97298-5_2

So a graph is a set of vertices equipped with an edge relation. We will be looking at visual representations of graphs and from those representations the graphs will inherit some additional features. Some vertices will be on the right, some on the left, some higher, some lower. Those are not the intrinsic properties of graphs, and they have to be disregarded. It will help to think of a graph as a database: just a list of vertices and edges.

Let us consider a graph, let's call it G, with the set of three vertices, which we will call a, b, and c. The edges of G are represented by ordered pairs (a, b), (b, a), and (c, c). The graph G has an edge between a and b, and a loop from c to c. There is a small technical inconvenience: the edge relation is symmetric, an edge between a and b is also an edge between b and a. Because relations are represented as sets of ordered pairs, this requires that for each edge between a vertex a and a vertex b, both pairs (a, b) and (b, a) are included in the edge relation of G. Here is a picture of the graph G:

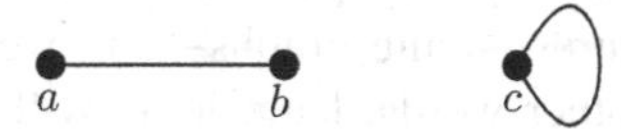

The picture of the graph G tells the whole story. Nothing is hidden, and much of what we see is formally expressible. For example, $\exists x \exists y[\neg(x = y) \wedge E(x, y)]$ expresses that G has an edge connecting two different vertices, and $\exists x E(x, x)$ that G has a loop. The sentence

$$\exists x E(x, x) \wedge \neg\{\exists x \exists y[\neg(x = y) \wedge E(x, x) \wedge E(y, y)]\}$$

expresses that G has only one loop.

For each graph G, by V_G we will denote the set of vertices of G and by E_G the set of ordered pairs representing the edges of V_G. Consider two graphs G and H each with only one vertex, as in the picture blow. V_G has one vertex a, and E_G is empty (no edges); V_H has one vertex b, and E_H has one edge, the loop (b, b).

The graphs are different, we see it instantly, and first-order logic detects this difference as well. Consider the sentence $\exists x \ E(x, x)$. It is true when interpreted in H and false in G.

Now let us consider two other graphs $\mathcal{G}$ and $\mathcal{H}$.[1] each with four vertices a, b, c and d. Let the set of edges of $\mathcal{G}$ be (a, b), (b, a), (b, c), (c, b), (c, d), (d, c) and the set of edges of $\mathcal{H}$ be (a, d), (d, a), (b, d), (d, b), (c, d), and (d, c).

Before we take a closer look at $\mathcal{G}$ and $\mathcal{H}$, let me stress an important point one more time. The illustrations are graphic representations of graphs. Those representations show all graph-theoretic features, but they also show more. We see points of a particular size, edges with a particular thickness. We see angles at which edges meet, we can measure distances between vertices. All those additional features are not features of the structures we want to analyze. The only information that matters is provided in the lists of vertices of edges. That information determines each graph uniquely, and we will see how much of that information can be expressed in first-order statements. For example, the statement $\forall x \forall y[E(x, y) \implies E(y, x)]$ is a first-order statement that is true in all graphs, because the edge relation is symmetric. Given a finite list of vertices, and a set of ordered pairs of vertices, one can verify that it is an edge relation, by checking that for every pair (a, b) in the set, the pair (b, a) is also there.

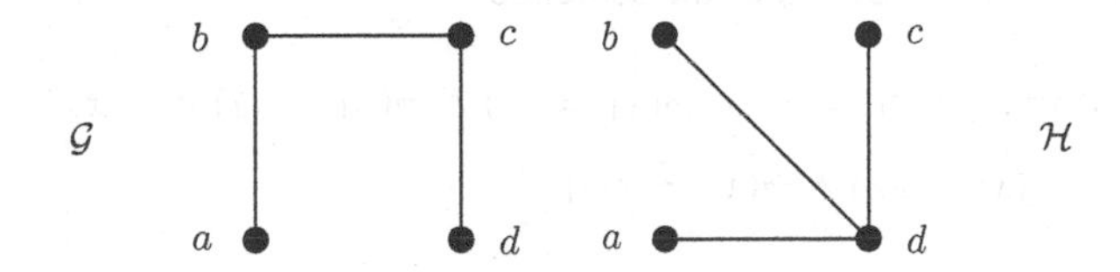

Graphs $\mathcal{G}$ and $\mathcal{H}$ are visibly different. How does logic see it? Consider the sentence

$$\exists x \forall y[\neg(x = y) \implies E(x, y)].$$

It says that there is a vertex connected by an edge to every other vertex. This sentence is true about $\mathcal{H}$ and false about $\mathcal{G}$. There are other first-order expressible properties that allow us to see the difference between $\mathcal{G}$ and $\mathcal{H}$ (think of some).

As graphs grow larger it becomes more difficult to spot differences between them just by looking, and so it is with logic, sentences expressing differences are becoming more complex. A commonly used measure of complexity of a sentence is the number of quantifiers it has. In our first example we could logically see the difference between G and H with the aid of a sentence with just one quantifier. In the second example the statement that expresses the difference between $\mathcal{G}$ and $\mathcal{H}$ has two quantifiers. It can be shown, and it is a consequence of a general theorem about first-order logic, that for any first-order sentence φ with only one quantifier, if φ is a

[1]Notice the change of font. Throughout the book we will discuss various mathematical objects, and for brevity, we give them "names" that are typically just single letters, sometimes using different fonts. There is nothing formal about those names, and choices of names are quite arbitrary.

true statement about $\mathcal{G}$ if and only if it is true about $\mathcal{H}$. It is not that surprising once one realizes that not much can be expressed about a graph by a sentence with only one quantifier. Think about it and try to come up with such sentences. For example, here is one $\exists x E(x, x)$. Seeing the difference between $\mathcal{G}$ and $\mathcal{H}$ is harder than seeing the difference between G and H. How hard can it get in general?

In graph theory, K_n denotes a complete graph on n vertices, i.e. the graph that has n vertices and every two vertices are connected with an edge. Here are picture showing K_3 and K_4:

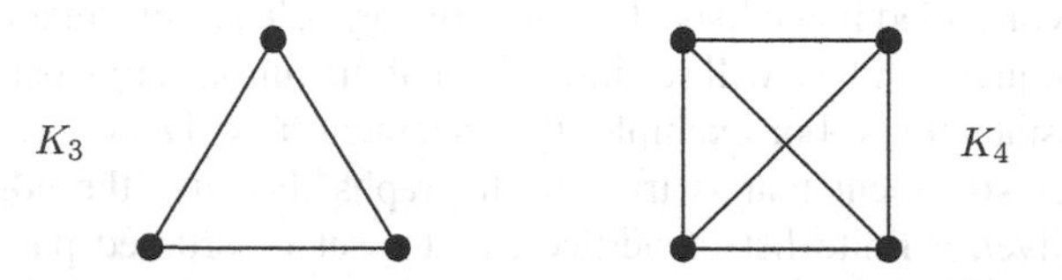

K_3 and K_4 are clearly different, but now it is a bit harder to find a first-order sentence that describes this difference. In fact, to see the difference in logical terms, we have to resort to a crude idea. Consider the sentence

$$\exists x_1 \exists x_2 \exists x_3 \exists x_4 [\neg(x_1 = x_2) \wedge \neg(x_1 = x_3) \wedge \neg(x_1 = x_4) \wedge \neg(x_2 = x_3) \wedge \\ \neg(x_2 = x_4) \wedge \neg(x_3 = x_4)].$$

This sentence says that there are at least four vertices, so it is true about K_4, but false about K_3. It is a crude sentence. It does not mention anything about the edge relation. It is hard to tell the difference between K_3 and K_4 in any other way, because the edge relations in both graphs are very similar; every vertex is connected by an edge to any other vertex. Nothing special happens there. The sentence witnessing the difference has four quantifiers, so it is rather complex. It follows from the same general theorem we mentioned before that the difference between K_3 and K_4 cannot be expressed by a sentence with fewer than four quantifiers. Now, given a visualization, we see the difference immediately, but that is because triangles and squares are the kind of shapes we can recognize instantly. A quick glance reveals that K_3 is a complete graph with fewer vertices than K_4, and that K_4 is also complete. But how can one see the difference between graphs such as K_{10} and K_{11}? The graphs are similar; every vertex is connected to any other vertex, and except that K_{11} has visibly more edges than K_{10} it is hard to describe any other difference. In fact, K_{10} and K_{11} share all properties that can be expressed by a first-order sentence with fewer than 11 quantifiers. In a sense, the only difference between K_{10} and K_{11} is their size. Logic can express this difference by a sentence with 11 quantifiers, and we cannot do much better. Given two large complete graphs with similar sizes, we cannot tell whether they are different or not without counting their vertices.

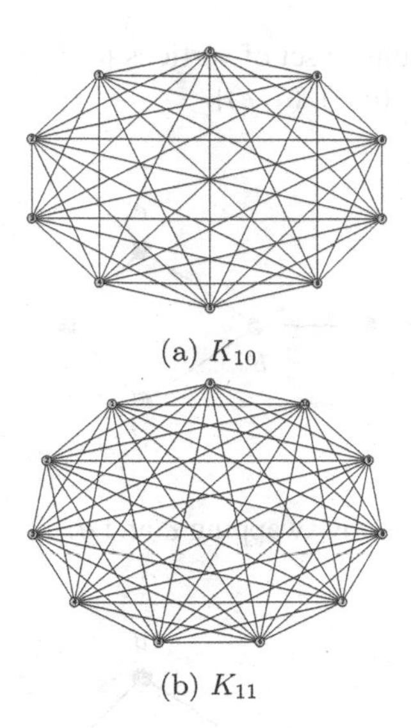

(a) K_{10}

(b) K_{11}

2.2 Symmetry

Symmetry is a subtle concept. In visual arts, music, and poetry it is used in many either straightforward or nuanced ways. Sometimes it is a desired effect, sometimes a feature to avoid. In classical geometry, the concept of symmetry is related to invariance under transformations. Rotating a square about its center by 90° results in an identical picture. If you think of a square as a collection of points, under such a rotation all points of the square have moved, but the picture looks the same as before. This is why the rotation by 90° is a symmetry of the square, but the one by 45° is not. In contrast, every rotation of a circle about its center is its symmetry. In this sense, the circle is more symmetric than the square. This example illustrates an important theme: We can study mathematical structures by investigating their symmetries. Knowing all symmetries of a structure tells us a lot about it, and in interesting cases it actually can characterize a structure completely.

Squares and circles are geometric objects, and what determines whether a certain transformation is a symmetry or not is whether the picture "looks" the same before and after. A similar idea can be applied to all mathematical structures, but we need to be precise about what is meant by "looking the same" before and after. What can change and what remains fixed? Logical seeing will help here.

Caveat: We will now discuss first-order properties of graphs that are easy to visualize, but the discussion requires a higher level of precision. We will talk about functions and we will use mathematical notation you may not be familiar with.

Since we represent edges as ordered pairs of vertices, if the pair (v, w) is in the set of edges of a graph, ten the pair (w, v) must be there as well. To save space, in the lists of edges of the graphs below we will list each pair only once with understanding that its symmetric counterpart is in the set as well.

Consider the graph G with the set of vertices $\{a, b, c, d, e\}$, and the set of edges $E_G = \{(a, b), (b, c), (b, d), (c, e), (d, e)\}$.

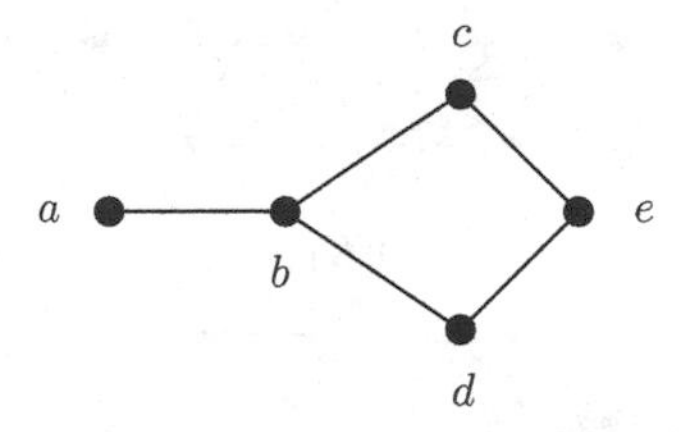

Let us rearrange the picture by swapping c and d.

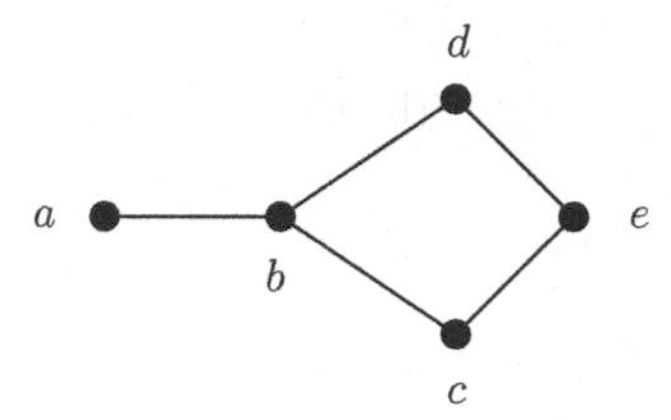

The image has changed. The vertex c which was up in the first picture, now is down, but what changed is not a first-order property of our structure. The picture representing the graph is now different, but it represents the same graph. Up, and down, are not "visible" without a geometric representation. The set of edges corresponding to the second picture, is the same as before. In other words, some vertices have moved, but the edge relation remained the same. Our transformation is a symmetry.

What can we learn about a structure by studying its symmetries? Let us examine our example more. Exchanging c with d was an obvious choice for the symmetry. Are there any other choices? A quick inspection shows that there are none. For example, exchanging a with b, is not a symmetry, and that is because a is connected by an edge only to one vertex, while b is connecter to three. If we swap a and b, the result will be a new graph G' whose edge relation is $E_{G'} = \{(a, b), (a, c), (a, d), (c, e), (d, e)\}$, so it is not the same as E_G. We can see the difference by looking at representations. How does logic see it? The notion of degree of a vertex will be helpful.

The *degree of a vertex* v is the number of vertices connected to v by an edge. In our graph G, the degree of a is 1, the degree of b is 3, and c, d, and e have degree 2. The formula

$$\exists y[E(x, y) \wedge (\forall z(E(x, z) \Longrightarrow z = y))]$$

expresses that the degree of a vertex x is 1. It says that there is a vertex y that is connected to x by an edge, and that there is only one such y; which is expressed by saying that if a vertex z forms an edge with x, then that z must be y. Let us call this formula $\varphi_1(x)$. When x is interpreted in G by a, $\varphi_1(x)$ becomes a true statement, and it is false when x is interpreted by any other vertex.

A small digression: As with some first-order statements we considered before, $\varphi_1(x)$ expresses the property we have in mind, but does it not in the most natural way. We have an advantage over logic, because we see the structure in its totality in its representation. We see that a is connected to b and to no other vertex. Logic does not have access to representations, it only knows the relation E_G as a set of ordered pairs. Given the set E_G, to verify that the degree of a is 1, one has to search the entire set and check that there is an edge (a, v) for some vertex v, and that there are no other edges (a, w), and this is really what $\varphi_1(x)$ expresses.

Here is a formula $\varphi_2(x)$ expressing that the degree of a vertex is 2, since we need more variables, we will be indexing them:

$$\exists x_1 \exists x_2[\neg(x_1 = x_2) \wedge E(x, x_1) \wedge E(x, x_2) \wedge (\forall y(E(x, y) \Longrightarrow (y = x_1 \vee y = x_2))].$$

For each natural number n there is a first-order formula φ_n expressing that the degree of a vertex is n. Thus, logic can see degrees of all vertices, and this gives us a necessary condition for a transformation to be a symmetry. Any symmetry has to map vertices of a given degree to vertices of the same degree. In the graph G, a is the only vertex of degree 1, and b is the only vertex of degree 3, neither a nor b can be moved by a symmetry. The other three vertices have degree 2, and we already saw that swapping c and d, while fixing all other elements is a symmetry. But how about swapping c and e? A quick glance at the representation of G shows that this is not a symmetry. We see it, and logic sees it too. Here is how. Consider the formula $\varphi(x)$ in which for brevity we use $\varphi_1(x)$ defined above:

$$\exists x_1 \exists x_2[E(x, x_1) \wedge E(x_1, x_2) \wedge \varphi_1(x_2)].$$

The formula expresses that in the graph one can get from the vertex x to a vertex of degree 1 moving along two edges. The formula is true when interpreted by c, but false when interpreted by e. In other words, $\varphi(x)$ is a first-order property that c has, but e does not. This is why swapping c and e is not a symmetry. That it is not a symmetry can also be checked by a direct inspection of the two sets of edges. Let H be the graph obtained by swapping c and e (draw a picture). Edge (b, e) is in E_H, but it was not in E_G. The edge relation has changed, so this transformation is not a symmetry.

We will now formally define the notion of symmetry of a graph, but first we need to fix some terminology. Above, we used the terms "transformations" and "mappings" of vertices without defining them. Now we will talk about functions. Functions are set-theoretic objects, and they will be introduced properly in Chap. 6. Informally, one can think of functions as operating on inputs, which in our case will be vertices of graphs, and turning out outputs, which will also be vertices. Functions are given names, for example f, and when we write $f(x) = y$ we mean that y is the output of f on the input x. Often, the inputs are called arguments, and the outputs are called values. To indicate that the inputs of a function f are all the elements of a set X and the outputs are all in the set Y, we write $f : X \longrightarrow Y$.

Functions can be defined by specifying a procedure for obtaining the output on given input, or they can be defined by a list of inputs and outputs. The transformation that turned out to be a symmetry in our first example above, is a function, let us call it f, that is defined by $f(c) = d$, $f(d) = c$, $f(a) = a$, $f(b) = b$, $f(e) = e$. A function is called *one-to-one* if does not assign the same output to distinct inputs. The function f above is one-to-one, if we altered it slightly by defining $f(a)$ to be d (for example) it would not be one-to-one.

A function $f : X \longrightarrow Y$ is *onto* if for every b in Y there is an a such that $f(a) = b$. If a function $f : X \longrightarrow X$ (not a typo) is one-to-one and onto, then we call it a *permutation* of the set X.

Definition 2.1 Let G be a graph. A permutation $f : V_G \longrightarrow V_G$ is a *symmetry* of G if and only if for any pair of vertices v and w,

$$(v, w) \text{ is in } E_G \text{ if and only if } (f(v), f(w)) \text{ is in } E_G. \qquad (*)$$

Notice that while $(*)$ in Definition 2.1 looks similar to the first-order formulas we have examined before, it is not a formula in the first-order language of graphs. The reason is that it involves the symbol f that is not part of the vocabulary. This statement is not about the graph G, rather, it is a statement about the graph G and the function f.

Definition 2.1 will be generalized later to arbitrary structures.

In the introduction to this section, we said that a symmetry is a transformation that moves elements of the structure, but does not change the way the structure "looks," and we tied the "look" of the structure to the first-order properties of its elements. Definition 2.1 does not refer to all first-order properties, but only to those that are expressible by the simplest formulas—the atomic ones. It would seem that by considering more properties in the definition, we could obtain a sharper, more restrictive notion of symmetry. This however is not the case, and it is due to the following theorem.

Theorem 2.1 *Let G be a graph. A permutation $f : V_G \longrightarrow V_G$ is a symmetry if and only if for every first-order formula $\varphi(x_1, x_2, \ldots, x_n)$ of the language of graphs and for any sequence of vertices $v_1, v_2, \ldots, v_n$*

$$\varphi(v_1, v_2, \ldots, v_n) \textit{ holds in } G, \textit{ if and only if } \varphi(f(v_1), f(v_2), \ldots, f(v_n)) \textit{ holds in } G. \qquad (**)$$

Compare $(*)$ and $(**)$. The statements are similar. The former says that symmetries preserve the edge relation; the latter says that in fact symmetries preserve all first-order properties as well. Theorem 2.1 has important consequences. Once know that a one-to-one function preserves the edge relation, we also know that it preserves all first-order properties. Hence; it implies that to show that a certain function is not a symmetry, it is enough to find one first-order property that it does not preserve.

2.3 Types and Fixed Points

In the previous section, we learned that if f is a symmetry of a graph G, and $f(v) = w$, then the vertices v and w must have exactly the same first-order properties. For finite graphs, the converse to this statement is also true. It is a theorem and to formulate it neatly, let us introduce one more definition. Let v be a vertex of a graph G. The *type* of v in G is the collection of all first-order formulas $\varphi(x)$ in the language of graphs, which are true in G when x is interpreted by v. Knowing a structure means, in particular, knowing all types of its elements. The definition of type is straightforward, and it does not involve anything technically complicated—the type is a set of certain formulas. As a set, it is itself a mathematical object, but as with many other mathematical objects, one thing is to define them, another is to know them. By analyzing a structure we can establish that an element in it has this or that first-order property, but this is far from being able to fully describe the complete type of the element. Types are usually very complex.

Theorem 2.1 implies that if f is a symmetry of a graph G, then for every vertex v, the type of v and the type of $f(v)$ are the same. If the graph G is finite, then this statement has an interesting converse, that follows from Theorem A.1 that is discussed in the Appendix.

Theorem 2.2 *Suppose v and w are vertices of a finite graph G, and that the type of v and the type of w are the same. Then, there is a symmetry f of G such that $f(v) = w$.*

Theorem 2.2 has important consequences. A complete description of a structure must include a list of types of its elements, but those types are often not easy to classify. The theorem says that we can see how many different types of elements a structure has by analyzing its symmetries. If two vertices of a finite graph have the same type, then there is a symmetry mapping one to the other. It follows that if a vertex is fixed by all symmetries of the graph, then it must have a unique type. Identifying those unique vertices is an important stage in getting to know a structure. Let us illustrate this with a simple example.

For a natural number n, an *n-star* is a graph that has one distinguished vertex forming edges with n other vertices, and has no other vertices or edges. Let S_n denote an n-star graph. Here are pictures of S_3 and S_5.

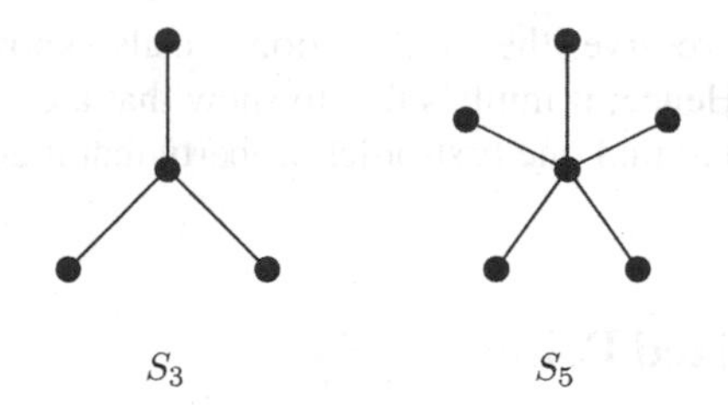

The stars S_1 and S_2 are exceptions, but beginning with $n = 3$, each graph S_n has vertices of exactly two types. The vertex in the center has degree n, and all other vertices have degree 1. Any permutation that fixes the center, is therefore a symmetry of S_n. Except for the center, a symmetry can swap any vertex with any other vertex. Thus we obtain a complete description of all symmetries of S_n. The center is unique, it is fixed by all symmetries. All other vertices are fungible.

2.4 Seeing Numbers

In the next chapter we will use first-order logic to build and analyze structures made of numbers. As discussed earlier, we will treat addition and multiplication as relations. As a relation, addition is the set of all ordered triples (k, l, m) such that $k + l = m$. Here is what this set looks like:

(0,0,0), (0,1,1), (0,2,2), (0,3,3), (0,4,4), (0,5,5), (0,6,6), (0,7,7), (0,8,8), (0,9,9), ...

(1,0,1), (1,1,2), (1,2,3), (1,3,4), (1,4,5), (1,5,6), (1,6,7), (1,7,8), (1,8,9), (1,9,10), ...

(2,0,2), (2,1,3), (2,2,4), (2,3,5), (2,4,6), (2,5,7), (2,6,8), (2,7,9), (2,8,10), (2,9,11), ...

(3,0,3), (3,1,4), (3,2,5), (3,3,6), (3,4,7), (3,5,8), (3,6,9), (3,7,10), (3,8,11), (3,9,12), ...

(4,0,4), (4,1,5), (4,2,6), (4,3,7), (4,4,8), (4,5,9), (4,6,10), (4,7,11), (4,8,12), (4,9,13), ...

(5,0,5), (5,1,6), (5,2,7), (5,3,8), (5,4,9), (5,5,10), (5,6,11), (5,7,12), (5,8,13), (5,9,14), ...

...

And here is multiplication:

(0,0,0), (0,1,0), (0,2,0), (0,3,0), (0,4,0), (0,5,0), (0,6,0), (0,7,0), (0,8,0), (0,9,0), ...

(1,0,0), (1,1,1), (1,2,2), (1,3,3), (1,4,4), (1,5,5), (1,6,6), (1,7,7), (1,8,8), (1,9,9), ...

(2,0,0), (2,1,2), (2,2,4), (2,3,6), (2,4,8), (2,5,10), (2,6,12), (2,7,14), (2,8,16), (2,9,18), ...

(3,0,0), (3,1,3), (3,2,6), (3,3,9), (3,4,12), (3,5,15), (3,6,18), (3,7,21), (3,8,24), ...

(4,0,0), (4,1,4), (4,2,8), (4,3,12), (4,4,16), (4,5,20), (4,6,24), (4,7,28), (4,8,32), ...

(5,0,0), (5,1,5), (5,2,10), (5,3,15), (5,4,20), (5,5,25), (5,6,30), (5,7,35), (5,8,40), ...

...

The addition and multiplication of natural numbers, as we will treat them in this book, are exactly as illustrated above. They are sets of ordered triples. How can we "see" this structure? How can we study it?

While addition and multiplication presented as relations are more complex than the comparatively simple graphs we considered in this chapter, there is much that both types of structures have in common. In the number structure instead of vertices, we have numbers, and instead of relations being given by sets of ordered pairs, they are given by sets of ordered triples, but the logical framework for studying properties is exactly the same.

Both sets of triples above are infinite, so we cannot see them in their totality, but we can inspect their finite fragments to look for special features and regularities. Also, there are some basic facts about both relations that can guide us. We are interested in identifying numbers, and sets of numbers with interesting properties, and in particular, we want to see which properties can be expressed by formulas of first-order logic.

Recall, that we designated A, M, and L, as relations symbols for addition, multiplication, and the ordering, respectively. Here are some examples of first-order properties written using those symbols.

- $A(x, x, x)$. There is only one number that has this property. It is 0, since $0+0=0$, and for all other numbers n, $n+n$ does not equal n.
- $M(x, x, x)$. There are exactly two numbers with that property: 0 and 1.
- $\exists y A(y, y, x)$. This formula defines the set of numbers that are of the form $y+y$. Those are exactly the numbers that ere divisible by 2, i.e. the even numbers.
- Here is a more complex property.

  $$\exists y \exists z[\neg M(y, y, y) \wedge \neg M(z, z, z) \wedge M(y, z, x)].$$

  In this formula, the only free variable is x. It defines the set of all those natural numbers x that have that property. The formula says that x is a product of two numbers y and z, and that y and z are neither 0 nor 1. In other words, it says that x is a composite number.
- And one more: $\exists y \exists z \exists v \exists w[M(y, y, v) \wedge M(z, z, w) \wedge A(v, w, x)]$. This is saying that for some y and z, $x = y^2+z^2$, so this formula defines the set of those natural numbers that can be written as a sum of two squares. For example, $0 = 0^2 + 0^2$, $1 = 0^2 + 1^2$, $4 = 0^2 + 2^2$, and $5 = 1^2 + 2^2$, so all those numbers have the property, but 3 does not.

The examples above show how some natural number theoretic properties can be expressed in first-order logic. I call such properties *logically visible*. In the language of first-order logic, using only the relation symbols for addition and multiplication, one can define almost all properties of natural numbers that number theorists are interested in. This is due to a special features of the arithmetic of natural numbers. In other, similar structures, logic allows one to see much less. This is the theme of the second part of this book.

To show that a certain property is expressible in first-order logic, one has to come up with a formal definition. Sometimes, those definitions are more or less straightforward translations of their informal versions. Sometimes, they are arrived at laboriously in a process that requires deeper insight into their nature.

To show that a certain property is not logically visible is another matter. Here is where new mathematical ideas come to play, and an important role is played by symmetries. We will make all this precise in Part II.

Exercises

Exercise 2.1 *Write a formula $\varphi_0(x)$ in the first-order language of graphs expressing that the degree of a vertex x is 0, and $\varphi_3(x)$, expressing that the degree of x is 3.*

Exercise 2.2 *Write a general form a formula $\varphi_n(x)$ in the first-order language of graphs expressing that the degree of a vertex is n, for each natural number $n > 0$.*

Exercise 2.3 *The random graph is an interesting mathematical structure. It has infinitely many vertices and is defined by the following property: for any two finite disjoint sets of vertices A and B, there is a vertex v that forms an edge with every vertex in A and none of the vertices in B. This property cannot be expressed by a single first-order sentence, but can be expressed by infinitely many. For each n, there is one sentence for sets A and B of size n. For example, the sentence for $n = 1$ is:*

$$\forall x_1 \forall x_2[\neg(x_1 = x_2) \Longrightarrow \exists x(E(x_1, x) \wedge \neg E(x_2, x)].$$

Write the sentences for $n = 2$ and $n = 3$.

Exercise 2.4 *Let V be $\{a, b, c\}$, and let E be $\{(a, b), (b, a), (b, c), (c, b), (a, a)\}$. Prove that the graph $G = (V, G)$ is rigid, i.e. it has no nontrivial symmetries.*[2]

Exercise 2.5 *Show that the following properties of natural numbers have first-order definitions in the language with relation symbols for addition and multiplication:*

1. *x is divisible by 3.*
2. *x is a sum of three squares.*
3. *x is a product of two prime numbers.*

[2]Every graph has the trivial symmetry, i.e. the symmetry f, such that $f(v) = v$ for each vertex v.

4. *x and $x + 2$ are prime (remember that 2 is not a symbol of first-order logic, so you cannot use 2 explicitly).*

Exercise 2.6 *Find a first-order property using a relation symbol for addition of natural numbers, that 2 has, but 3 does not.*

Chapter 3
What Is a Number?

> *In a scientific technique there is almost always an arbitrary element, and a philosophical discussion which puts too much stress on the 'technical' aspects of the problem in question, exposes itself all too easily to the suspicion of resting for a part on purely arbitrary stipulations.*
>
> Evert W. Beth, *Aspects of Modern Logic* [5].

Abstract In Chap. 1, we used addition and multiplication of the natural numbers to introduce first-order logic. Now, equipped with formal logic, we will go back and we will *reconstruct* the natural numbers and other number systems that are built on them. This looks circular, and to some extent it is. The set of natural numbers with a set of two relations—addition and multiplication—is a fundamental mathematical structure. In the previous discussion, we took the structure of natural numbers for granted, and we saw how some of its features can be described using first-order logic. Now we will examine the notion of natural number more carefully. It will not be as easy as one could expect.

Keywords Natural numbers · Arithmetic operations · Decimal system · Zero

3.1 How Natural Are the Natural Numbers?

As often happens, the most common ideas turn out to be hardest to pinpoint exactly. In "New Athens," the first Polish encyclopedia, published in 1745, Benedykt Chmielowski wrote "What a horse is, everyone can see, and goats—smelly kind of animal."[1] Nowadays, Wikipedia informs us that "The horse (Equus ferus caballus) is one of two extant subspecies of Equus ferus. It is an odd-toed ungulate mammal belonging to the taxonomic family Equidae." Horses are complex creatures, but they can be given a more or less precise definition. But what is a number? According

[1] My translation.

R. Kossak, *Mathematical Logic*, Springer Graduate Texts in Philosophy 3,
https://doi.org/10.1007/978-3-319-97298-5_3

to Google, it is "an arithmetical value, expressed by a word, symbol, or figure, representing a particular quantity and used in counting and making calculations and for showing order in a series or for identification." This looks fine until we ask: What is arithmetical value, or What is quantity? We read further: "A number is a mathematical object used to count, label, and measure." So here we have even more of a challenge, we need to explain what is a "mathematical object." Intuitive understanding is assumed. The encyclopedists seem to be saying: what is a number, everyone can see. This is perfectly fine for practical applications, but our discussion here is at a different level. Our goal is to understand mathematical structures. Since some of the most important structures are made of numbers, it would be good to know what those numbers are, so now I will outline a modern approach to numbers and number systems that was developed in nineteenth century by Georg Cantor, following important earlier work by Richard Dedekind.

Numeracy is a basic skill. We all know numbers, but keep in mind that most of us are familiar not with numbers as such, but rather with their representations. Think of a number, say 123. What is 123? It is a sequence of digits. To know what this sequence *represents*, we need to understand the decimal system. The symbols 1, 2, and 3 are digits. Digits represent the first ten counting numbers (starting with zero). The number corresponding to 123 is $1 \cdot 100 + 2 \cdot 10 + 3 \cdot 1$. In this representation, the number has been split into groups: three ones (units), two tens, and one hundred. The example illustrates how the decimal system works, but it does not explain what the number 123 is, and what natural numbers are in general.

There is nothing special about counting in groups of ten. If instead of groups of tens, we counted in groups of fives, the number 123 would be written as 443, since it is the sum of four twenty-fives, four fives, and three units. If we used the quinary system instead of the decimal, the powers of 5, $5^2 = 25$, $5^3 = 625$, would be known as 100, 1000, and they would have some special names.

The decimal system is very efficient, but the fact that we depend so much on it when we learn basic arithmetic strongly affects the common perception of numbers. To try to get to the essence of natural numbers, we will start with something more basic—counting.

What is counting? We look at a collection of objects and we count: one, two, three, four, ... In the search for basic insights, this is not getting us far, since counting this way presupposes numbers already. So let us forget about numbers for a moment. Notice that we can still count. How? One way would be to point at objects one by one, and say something like: one, one, one... . This is a rudimentary method, but it can be of some use. Looking at a herd of sheep and a collection of pebbles, we can count both of them this way simultaneously, and we can learn something. If we stop one count before the other, we will know whether there are more sheep than pebbles, or more pebbles than sheep. If we stop at the same time, we know that both sets have the same size. This conclusion will be our point of departure. We will not go into whether such counting can always be performed, or what kinds of collections it can be applied to, but let us agree that we do have common intuition about the process itself. We will try to build on that.

To arrive at the notion of number, we will begin with comparison of sizes. To perform such comparisons, one does not even have to count. There is another way. Think of two collections of pebbles. To compare their sizes, one can make two rows, one next to the other, with the same distance between consecutive pebbles. Once we see the rows, we know how the sizes compare. To make such comparisons, we need to be able to organize collections of elements in rows, or, as we will be saying, we need to be able to *order* them. In each such row, there will be a first element, a second element, and so on. We are not *using* numbers here yet. We just see the distinct element at the beginning of the row, then the next right behind it, and the next, and so on. We are not talking here about practical aspects of ordering, counting, and comparing. Our goal is to use a general intuition to motivate a concept.

Since we appeal only to basic intuitions, the discussion so far seems rather trivial, but with a bit more thought, a more sophisticated concept emerges. We can observe that among all rows are those that are the shortest, and any two such shortest rows are of the same size. Let us call this size *two*. If we adjoin one object to one of those smallest rows, we obtain a collection that is *immediately* larger. We call the size of this collection *three*. Next, we generalize. We see that for each size there is a size immediately larger, obtained by adding another element to a collection whose size is already established. Once we learn how to recognize some initial sizes, it is natural to given them names: one, two, three, ...

You may be concerned that we did not start naming sizes starting withe the size of a single object, and naming it *one*. We could have, but we are trying hard to stay close to absolutely clear and basic intuitions, the concept of *one* requires a more careful justification, and we will not do it here. Once we see how to extract the notion of number from comparisons of sizes of ordered collections, it is reasonable to accept also this extreme case of the shortest one one-element row, but it is an afterthought. This is a subtle point, and I am certainly not the first to bring it up. It was Edmund Husserl, who in his insightful analysis of psychological roots of arithmetic concepts in *Philosophy of Arithmetic: Psychological and Logical Investigations* [16], first published in 1891, argued that 1 is a *negative* number since it signifies absence of multiplicity. And what about zero? If we root our understanding of numbers in counting, then the concept of zero does not arise in a natural way, but we can think of it as the size of a row that is still waiting for its first element; it has not started forming yet.

The discussion above serves as an illustration of how a more abstract notion, that of number, is grounded in a more rudimentary one, that of counting. We will not go into what makes a concept rudimentary or what phenomenology would call "immediately given." For us it is enough that the concept does not, in any obvious way involve anything already formally mathematical. There are many ways in which such grounding of higher level concepts can be described, and there will always be a debate on whether the lofty goal of complete clarification has indeed been achieved. There is always room for doubt. In order to make progress, the grounding process has to stop somewhere. For us, it stops at the concept of counting. We will also assume for now that we know intuitively what a collection of objects is, and what it means to order it.

In our approach to numbers, one could suspect some trickery. The point was to explain the nature of natural numbers, and to do so we talked about sizes of collections. We identified a few initial small sizes and gave them names that clearly sounded like numbers. Here again appears to be some circularity. There is a subtle difference though. Numbers can serve as measures of sizes, but the idea of size is more basic. It is more basic in the sense that we do not need to assign any abstract quantities to collections to be able to *compare* their sizes. We can see whether or not two collections are of the same size, without knowing the number of their elements.

In the process of grounding we can go a bit further. To arrive at the concept of number, we do not even need the notion of size. Instead, we can just rely on an intuitive understanding of *larger* and *smaller*. Then, the notion of size emerges as follows. Consider a collection of elements. We can compare this collection to other collections, as described above. Some collections will be smaller, some larger. For some collections, the simultaneous counting process will terminate at the same time. If this happens for two collections, then they are in a certain precise sense equivalent. To identify this sort of equivalence we will use the term *equinumerous*. For any collection, there is a multitude of collections equinumerous with it. The collection of fingers of my right hand is equinumerous with the collection of fingers of my left, and they both are equinumerous with the collection of the USB slots on my laptop, and equinumerous with the collection of all oceans. All those equinumerous collections have their individual features, but there is one thing that they all have in common. That is this *one thing* that we call the size. This common feature *is* the size of the collection and of all other collections that are equinumerous with it.

Now we can introduce the following, more formal definition: a *counting number* is the size of a finite collection. In the definition, the new concept of a counting number is defined in terms of concepts already clarified (or assumed to be clear), with an important exception, we have not yet said a word about what it means for a collection to be finite. One is tempted to say that a collection is finite if it has a finite number of elements, but that is an obvious circularity. In the discussion so far, we tacitly assumed that all collections were finite in the common understanding of this word. In fact, the qualification "finite" is needed only if one anticipates collections that may not be finite. Allowing infinite collections in the discussion is a big step towards abstraction. Up to this point, we could have assumed that all collections were finite, but soon we will be very much concerned with infinite collections. It is the admittance of infinite mathematical structures that gave a great boost to the developments that are the subject of this book. It turns out that the discussion of size, as we described it above, can be generalized to infinite collections in a precise way. It makes sense to talk about the size of any collection, and there is an interesting arithmetic of infinite numbers associated with those sizes. We will say more about it in Chap. 6.

3.1.1 *Arithmetic Operations and the Decimal System*

The decimal system assigns symbolic representations to all counting numbers. It also provides practical recipes for addition and multiplication. For example

$$\begin{array}{r} 123 \\ +88 \\ \hline 211 \end{array}$$

Given symbolic representations of two or more numbers, we can, as in the example above, mechanically perform algorithmic operations on their digits and obtain a symbolic representation of their sums. The familiar algorithms for addition and multiplication do not operate on numbers, they operate on strings of digits, but addition and multiplication as such are defined independently of any system of representations. How? Let us go back to the definition of counting numbers. What is three plus two? Think of an ordered set of size three extended by a set of size two in such a way that all elements of the second set come after the elements of the first. The size of the ordered set thus obtained is the result:

$$\bullet\ \bullet\ \bullet + \bullet\ \bullet = \bullet\ \bullet\ \bullet\ \bullet\ \bullet$$

Multiplication is repeated addition. Three times two is $2 + 2 + 2$. To define addition of natural numbers, we employed counting. For multiplication, we need another concept—repetition. Is repetition an everyday notion that needs no formal clarification or are we importing a concept of mathematical nature? This calls for a deeper philosophical analysis, but, instead, we will use a different approach. Instead of repetition, we will invoke another basic idea, that of splitting or breaking into pieces. If three pebbles are each broken into two pieces, the result is six smaller pebbles. The pebbles have multiplied. See the illustration below.

$$\bullet\ \bullet\ \bullet \times \bullet\ \bullet = \begin{array}{c} \bullet\ \bullet\ \bullet \\ \bullet\ \bullet\ \bullet \end{array} = \bullet\ \bullet\ \bullet\ \bullet\ \bullet\ \bullet$$

This is a set theoretic way of introducing multiplication. It allows to generalize the notion to the case of infinite numbers in a natural way. If m and n are natural numbers, then the product $m \cdot n$ can be defined to be the size of the set obtained from a collection of m elements, by replacing each element with a collection of size n. Notice that the concept of multiplication is defined without any reference to number representations.

3.1.2 How Many Numbers Are There?

In practice, there are natural limitations on the sizes of sets that can actually be counted. We live in a bounded region of the physical space, and our lives are not endless, but even regardless the human aspect, in modern physics the knowable universe is described as huge, but still finite collection of elementary particles, so, in this sense, it is finite. But could there be a largest number? This does not seem to make much sense. If we can count up to a number n, surely we can count up to the next number $n + 1$. The fact that we accept that there is no largest number has profound consequences. It forces us to consider potentially infinite processes, and hints at the necessity to consider mathematical objects that are actually infinite.

Thinking in the other direction, one can imagine a number system without infinity. For example, in a society in which sizes 1, 2, 3, and 4 are recognized, and each collection of more than four elements is considered large, counting and the number system would be simple: one, two, three, four, many. There still would be some general mathematical rules, such as $m + n = m + n$, or many + many = many, but there would be no need to invoke any concept of infinity.

3.1.3 Zero

In the discussion above the size of a smallest size was called one. In a move towards abstraction, we can introduce the concept of *empty* collection. An empty collection is a collection that has no elements, for example the collection of chairs in an empty room. An empty collection is not nothing. It is an object. The size of an empty collection is called zero. Thus we extended the domain of counting numbers, by adding the number zero. It follows from the definitions of addition and multiplication we gave above, that for each number n, $n + 0 = n$ and $0 \cdot n = 0$. While it is clear that if we consider multiplication as repeated addition, then $n \cdot 0$ is the size of the n empty sets put together, hence it is zero. To justify that $0 \cdot n = 0$ using the interpretation that involves splitting, one can say that since there are no elements to split in an empty set, nothing has been added to the set; hence its size remains 0.

Whether or not to include zero as a counting number is a philosophical question, but for practical applications and further developments in algebra, the concept of zero is crucial. For example, in the decimal system 0 is a placeholder marking the absence of certain powers of 10 in the decimal representation of a number. For example, 2017 denotes the size of a set that can be split into $2 \cdot 1,000 + 0 \cdot 100 + 1 \cdot 10 + 7 \cdot 1$. There are no hundreds. The collection of hundreds in the decimal representation of 2017 is empty.

3.1.4 *The Set of Natural Numbers*

In mathematics, the counting numbers are called *natural*. Also, instead of the term "collection" we will more often use the word *set*. The set of natural numbers is usually denoted by $\mathbb{N}$ and we will follow this convention. In other words, $\mathbb{N} = \{0, 1, 2, \dots\}$.[2]

In the last sentence above, we did something profound. We declared that the symbol $\mathbb{N}$ denotes a mathematical object and, as indicated by the dots, that object is not finite. The set $\mathbb{N}$ is *actually infinite*. We could have said, the natural numbers are: $0, 1, 2, \dots$, having in mind a never-ending process in which for each natural number we can find or construct a number that is one larger. That would involve the idea of *potential infinity* —a process that has no end. Instead we wrote $\mathbb{N} = \{0, 1, 2, \dots\}$. Just by adding two curly brackets, we brought to life a new kind of object—an infinite set. Formal set theory which we will discuss soon, provides a framework for such acts of creation.

Exercises

The arithmetic of natural numbers is governed by a set of simple rules. Addition is *commutative* and *associative*, which means that for all natural numbers a, b, and c, $a + b = b + a$, and $a + (b + c) = (a + b) + c$. Based on the direct intuition of addition, these two rules can be easily justified. Multiplication is also commutative and associative, but how do we know that? For example $3 \cdot 2 = 2 + 2 + 2 = 6$, and $2 \cdot 3 = 3 + 3 = 6$, but how do we know that it is always the case? In the following exercises, you are asked to provide convincing justifications for the following statements. Hint: Think about rectangles and boxes.

Exercise 3.1 *Explain why, for all natural numbers a and b, $a \cdot b = b \cdot a$.*

Exercise 3.2 *Explain why, for all natural numbers a, b,and c, $a \cdot (b \cdot c) = (a \cdot b) \cdot c$.*

Exercise 3.3 *Explain why, for all natural numbers a, b,and c, $a \cdot (b + c) = (a \cdot b) + (a \cdot c)$. Hint: Draw a rectangle with on side of length a and the other of length $b + c$.*

[2]In the set-theoretic tradition, I included 0 in the set of all natural numbers. Another popular convention, adopted in many textbooks, is to start natural numbers with 1, and then to call the set $\mathbb{N}$ defined above either the set of whole numbers, or the set of nonnegative integers.

Chapter 4
Seeing the Number Structures

I thought, Prime Number. A positive integer not divisible. But what was the rest of it? What else about primes? What else about integers?

Don DeLillo *Zero K* [7]

Abstract In the previous chapter, we introduced and named an actually infinite set. The set of natural numbers $\mathbb{N} = \{0, 1, 2, \dots\}$. What is the *structure* of this set? We will give a simple answer to this question, and then we will proceed with a reconstruction of the arithmetic structures of the integers and the rational numbers in terms of first-order logic. The reconstruction is technical and rather tedious, but it serves as a good example of how some mathematical structures can bee seen with the eyes of logic inside other structures. This chapter can be skipped on the first reading, but it should not be forgotten.

Keywords Set notation · Linearly ordered sets · Integers · Fractions · Rational numbers · Equivalence relations · Densely ordered sets

4.1 What Is the Structure of the Natural Numbers?

From our point of view, the question in the title of this section is ill-posed. This whole book is about mathematical structures understood in a specific way. To study a mathematical structure one needs to define it first. A structure is a set with a set of relations on it. We have the set $\mathbb{N}$, but we have not yet defined any relations. The set $\mathbb{N}$ itself comes with an empty set of relations, and as such is not different (as a structure) from any other infinite set of the same size with no relations on it.[1] It is up to us what relations we want to consider. There is a vast supply of relations, and they can be put together to form many structures.

[1]We have to wait until Chap. 6 for a discussion of sizes of infinite sets.

R. Kossak, *Mathematical Logic*, Springer Graduate Texts in Philosophy 3,
https://doi.org/10.1007/978-3-319-97298-5_4

Typically, in mathematics, a relation is defined by a rule that allows one to verify whether given objects are related or not. Often, but not always, it is given a special name. Here is an example that I just made up: say that natural numbers m and n are *equidigital* if their decimal representations have the same number of different digits. According to this rule, 123123 and 987 are equidigital, and 2334 and 1234 are not. Equidigitality is a perfectly well-defined relation. As an exercise you may want to define a few similar relations on your own. Some such relations may be of special mathematical interest, most are meaningless. In this chapter we will examine some basic relations that are essential in the study of number structures. We will talk about the ordering, addition, and multiplication of numbers, but before we get into details, I need to introduce more notation.

4.1.1 Sets and Set Notation

Since most mathematical objects can be considered as sets, we need to say more about what sets are, and we will need some notation to do that. Initially, one can think of sets as just collections of objects, but not much is gained by replacing the word "set" by the word "collection." The question, "What is a set?" has many answers. Georg Cantor who created set theory in an effort to generalize concepts of counting and number beyond the finite, wrote, "A set is a Many that allows itself to be thought of as a One." What it could mean requires a thorough discussion, but for now we only need to introduce some notational conventions.

Consider the set of letters in the word "set." Let us call it S. We can also say, let $S = \{s, e, t\}$. We could also define the same set by saying that $S = \{e, s, t\}$ because in sets the order of elements is irrelevant. Another way to define the same set S is

$$S = \{x : x \text{ is a letter in the word "set"}\}.$$

Here we think of x as a variable in some formula expressing a property of unspecified objects of some kind. Often this formula will be written in a formal language, but not always, as is the case of this definition of S.

In general, $\{x : \varphi(x)\}$ is the set of elements x which have a certain property $\varphi(x)$. If a set is small, we can just list all its elements. For larger sets, the last form of the definition above is most convenient.

The elements of a set are also called its members. Thus, in the example above, e is a member of the set S we defined above, but a is not. The commonly used symbol for the membership relation is $\in$. In symbols: $e \in S$, and $a \notin S$.

Since we pay much attention to the separation of syntax and semantics, a caveat is needed. The set theoretic notation just introduced is not a part of the formalism of first-order logic. We use this notation and symbols to abbreviate mathematical statements. Later, we will discuss formal theories of sets, and then the membership relation symbol $\in$ will be also used for a relation symbol of first-order logic.

4.1.2 Language of Formal Arithmetic

In the chapter on first-order logic we introduced three relation symbols L, A, and M. The symbol L was interpreted in the domain of natural numbers as the relation "less than," and A and M as addition and multiplication both considered as relations. Using the set notation and logical symbols, L was interpreted as the set

$$\{(k, l) : k \in \mathbb{N} \land l \in \mathbb{N} \land k < l\},$$

A as the set

$$\{(k, l, m) : k \in \mathbb{N} \land l \in \mathbb{N} \land m \in \mathbb{N} \land k + l = m\},$$

and M as the set

$$\{(k, l, m) : k \in \mathbb{N} \land l \in \mathbb{N} \land m \in \mathbb{N} \land k \cdot l = m\}.$$

We will call those sets Less, Add, and Mult, respectively. For example (0, 0, 0) and (1, 2, 3) are in Add, but (1, 1, 1) is not.

Let me stress one more time that while L, A, and M are formal relation symbols of first-order logic, the names Less, Add, and Mult are informal mathematical abbreviations.

The three relations can be used in different configurations to form different structures. We will discuss four of them: the natural numbers with the ordering relation $(\mathbb{N}, \text{Less})$, the additive structure $(\mathbb{N}, \text{Add})$, the multiplicative structure $(\mathbb{N}, \text{Mult})$, and the most interesting $(\mathbb{N}, \text{Add}, \text{Mult})$ which we will call the *standard model of arithmetic*.

4.1.3 Linearly Ordered Sets

One does not need advanced mathematics to analyze $(\mathbb{N}, \text{Less})$. The picture below shows what it looks like. It is a sequence with a first element on the left, and it extends without a bound to the right. It is an infinitely long queue, and it is not important that it is illustrated as beginning on the left and going to the right, although this is how it is usually thought of, because we like to see the numbers on the number line as increasing from left to right. In $(\mathbb{N}, \text{Less})$, it is not important that the elements of the set $\mathbb{N}$ are numbers. It only matters that the set $\mathbb{N}$ is infinite and that the relation Less orders its elements in a particular way.

The picture below tells us the whole story.

● ● ● ● ● ...

We will use ($\mathbb{N}$, Less) to illustrate an approach that we are going to apply later to more complex structures.

For now, we will consider first-order logic in the language with only one binary relation symbol L. We will start with a formal definition involving three properties formalized in this language. Those statements are about structures of the forms (M, R), where M is a set, and R is a binary relation on M. In other words, we interpret the symbol L, as the relation R on M. The meaning of the three conditions is explained right after the definition.

Definition 4.1 Let R be a binary relation on a set M. We say that R *linearly orders* M, or that R is a *linear ordering*, if the following three statements are true in (M, R), when L is interpreted as R:

(O1) $\forall x \neg L(x, x)$.
(O2) $\forall x \forall y[(x = y) \vee L(x, y) \vee L(y, x)]$.
(O3) $\forall x \forall y \forall z[(L(x, y) \wedge L(y, z)) \Longrightarrow L(x, z)]$.

Think of the elements of the set M forming a line, like the natural numbers in the illustration of ($\mathbb{N}$, Less) above. The idea here is that the line is determined by the relation R in the sense that a stands before b if and only if the pair (a, b) is in the set R. With this in mind, you can see that *O1* says that no element stands before itself; *O2* says for any two elements a and b, either a stands before b, or b before a; and *O3* says that if a stands before b and b before c, then a is before c as well.

Clearly, ($\mathbb{N}$, Less) has all three properties, hence Less linearly orders $\mathbb{N}$. We will see other examples of linearly ordered sets later, but for just one example now, let us note that if we define the relation More to be the set of pairs (a, b) such that (b, a) is in Less, then More also linearly orders $\mathbb{N}$. The structure ($\mathbb{N}$, More) has a largest element, and no least element.

4.1.4 The Ordering of the Natural Numbers

The relation Less linearly orders $\mathbb{N}$, but ($\mathbb{N}$, Less) also has a number other of properties that can be expressed by the first-order statements that distinguish it from other linearly ordered sets. Here are some of those properties with the translations to the informal language provided below.

(1) $\exists m \forall n\ [(n = m) \vee L(m, n)]$.
(2) $\forall m \exists n\ L(m, n)$.
(3) $\forall m \exists n\{L(m, n) \wedge \forall k[L(k, n) \Longrightarrow ((k = m) \vee L(k, m))]\}$.

Here are the translations:

(1) There is a least natural number.
(2) There no greatest natural number.
(3) Every natural number has an immediate successor (see below for an explanation).

A linear ordering is *discrete* if for every element a, except for the last element if there is one, there is a b that is larger than a, and there are no elements between a and b. We call such a b an immediate successor of a.

The combined content of the statements *O1*, *O2*, *O3*, and (1), (2), (3) above, can be summarized with just one sentence: Less is a discrete linear ordering with a least element and no last element. That one sentence tells us almost everything there is to know about ($\mathbb{N}$, Less). This is made precise by the following theorem. For a proof of a similar result see Proposition 2.4.10 in [22].

Theorem 4.1 *If a binary relation* R *is a discrete linear ordering of a set* M *with a least element, and every element has an immediate successor, and every element, except the least one, has an immediate predecessor, then a first-order sentence* φ *is true in* (M, R) *if and only if it is true in* ($\mathbb{N}$, Less).

Theorem 4.1 says that (M, R) and ($\mathbb{N}$, Less) have exactly the same first-order properties. However, it does not preclude that there may be differences between (M, R) and ($\mathbb{N}$, Less), and if there are such differences, that means that they cannot be expressed in the first-order way. Let us see an example.

Here is a picture of a set M linearly ordered by a relation R, the ordering is from left to right:

$$\bullet\bullet\bullet\bullet\bullet\cdots \qquad \cdots\bullet\bullet\bullet\bullet\bullet\cdots$$

The idea here is that (M, R) has a part (the one on the left) that looks like ($\mathbb{N}$, Less) and each element in it is smaller than all elements from the rest of the structure that is a discretely ordered set that stretches to infinity in both directions. It is easy to see that the statements (1), (2), and (3) are true about (M, R); hence, by Theorem 4.1 (M, R) is first-order indistinguishable from ($\mathbb{N}$, Less). There is a clear difference though. The domain of new structure has a subset without a least element (the infinite part on the right). There are no such subsets of $\mathbb{N}$. It follows that this difference is not first-order expressible.

4.2 The Arithmetic Structure of the Natural Numbers

We defined the addition relation Add on the set of natural numbers as the set of ordered triples (k, l, m) such that $k + l = m$. Think of the set Add as an oracle that knows correct results of all additions. To find out whether 253 plus 875 is 1274, there is no need to perform any operations. Just ask the oracle if the triple (253, 875, 1274) is in Add (the answer is "no"). But this is all an oracle that knows a relation can do. If we want to know what is $253 + 875$, we ca still use the oracle, but we cannot ask it for the answer directly. We need to ask all questions of the form $(253, 875, \star)$ for different values of $\star$ until we get a "yes." To clarify, an oracle is a data base and it knows what it has, but it cannot automatically respond to queries of the kind that require more elaborate searches. It does not understand when we ask: Tell me what number is the third entry in the triple you have that starts with 253 and 875.

Recall that A is a ternary relation symbol, that we have already used to examine some of first-order statements about Add. Here are three more examples; their informal translations follow.

(1) $\exists x \forall y\, A(x, y, y)$.
(2) $\forall x \forall y \forall z\, [A(x, y, z) \Longrightarrow A(y, x, z)]$.
(3) $\forall x \forall y \forall w \forall z\, [(A(x, y, w) \wedge A(x, y, z)) \Longrightarrow w = z]$.

The first sentence declares the existence of a natural number with a special property. It is a number that is "neutral" with respect to addition. There is only one such number—zero. The second expresses that addition is commutative, meaning that the order in which the numbers are added does not matter. The third sentence does not seem very interesting, but it is of some importance. We chose to formalize statements about addition in a relational language, i.e. instead of introducing a function symbol for the operation of adding numbers, we represent addition as a relation. Sentence (3) expresses that the addition is in fact a function; for any two numbers, their sum is unique.

The multiplicative structure $(\mathbb{N}, \mathsf{Mult})$ is interesting and there will be more about it later, but now we will turn to an even more interesting structure in which addition and multiplication are combined. It is $(\mathbb{N}, \mathsf{Add}, \mathsf{Mult})$ which we called the *standard model of arithmetic*. We have seen before how certain number-theoretic facts can be expressed as first-order properties of $(\mathbb{N}, \mathsf{Add}, \mathsf{Mult})$. The examples included Euclid's theorem on the infinitude of prime numbers. In two exercises in Chap. 1, the reader was asked to formalize the Twin Primes Conjecture, and the Goldbach Conjecture. The fact that such statements can be written just in terms of addition and multiplication already shows how rich the standard model is and how much of this richness can be expressed in first-order logic. In fact, the standard model is one of the most complex structures studied in mathematics. There is much that we know about it, and even more that we don't. In contrast, the structures $(\mathbb{N}, \mathsf{Less})$, $(\mathbb{N}, \mathsf{Add})$, and $(\mathbb{N}, \mathsf{Mult})$ are relatively simple and well-understood. All of this will be discussed later in the book, but our immediate goal is first to introduce other number systems that are extensions of $(\mathbb{N}, \mathsf{Add}, \mathsf{Mult})$.

4.3 The Arithmetic Structure of the Integers

When we introduced the natural numbers, we appealed to the simple idea of counting. To introduce negative numbers, we will follow a formal route. We will start from the already introduced structure $(\mathbb{N}, \mathsf{Add}, \mathsf{Mult})$, and construct over it a system of integers (positive and negative whole numbers) it using first-order logic. The purpose is to show how new structures can be built form the ones we already have. Be prepared that the presentation will be more formal and not particularly intuitive. There are good reasons why the presentation cannot be too straightforward. Although technically there is nothing difficult about the basic rules of arithmetic of integers numbers, the historical development of the foundations of arithmetic was quite convoluted, and it took a while until the mathematical community came to a consensus. Morris Kline writes about it in [18], and he quotes Augustus de Morgan:

> The imaginary expression $\sqrt{-a}$ and the negative expression $-b$ have this resemblance, that either of them occurring as the solution of a problem indicates some inconsistency or absurdity. As far as real meaning is concerned, both are equally imaginary, since $0-a$ is as inconceivable as $\sqrt{-a}$, [18, p. 593].

Recall that the set on which the relations of a structure are defined is called its *domain*. Thus, the domain of $(\mathbb{N}, \mathsf{Add}, \mathsf{Mult})$ is $\mathbb{N}$. We will build a new structure with a domain $\mathbb{Z}$ and three relations on it: $\mathsf{Less}_{\mathbb{Z}}$, $\mathsf{Add}_{\mathbb{Z}}$ and $\mathsf{Mult}_{\mathbb{Z}}$.[2] Ultimately, $\mathbb{Z}$ will become a set of numbers, but initially its elements will not look like ordinary numbers at all.

Let $\mathbb{Z}$ be the set of all ordered pairs $(0, m)$ and $(1, m)$, where m is any natural number, except that we exclude $(0, 0)$. The idea here is that all pairs $(1, m)$ represent the nonnegative integers, and all pairs $(0, m)$ their negative opposites. Think of $(1, m)$ as the number m, and $(0, m)$, as its opposite $-m$. We have excluded $(0, 0)$ to avoid duplication. According to the definition, zero is represented by $(1, 0)$.

What is the purpose of defining integer numbers this way? The point is to seriously address the question: Where do mathematical objects come from? It is a difficult philosophical problem that does not have an ultimate answer. We have followed the route of grounding the concept of natural number in our intuitive grasp of counting. Then, we "made" the standard model $(\mathbb{N}, \mathsf{Add}, \mathsf{Mult})$, and now we are extending it to a larger number system of integers. We will continue extending it, culminating in the complex numbers, and we want to see what is this process based on, and how the newly constructed structures are justified. For a while no now tools will be needed. We will construct the integers and the rational numbers (fractions) by defining them in the standard model using first-order logic.

The set $\mathbb{Z}$ is first-order definable in $(\mathbb{N}, \mathsf{Add}, \mathsf{Mult})$. Here is a formal definition:

$$\mathbb{Z} = \{(i, j) : i \in \mathbb{N} \wedge j \in \mathbb{N} \wedge (i = 0 \vee i = 1) \wedge \neg(i = 0 \wedge j = 0)\}.$$

Since the domain $\mathbb{Z}$ is made of pairs of numbers, binary relations on it will be represented as a sets of pairs of pairs (not a typo) of numbers.

Definition 4.2 The relation $\mathsf{Less}_{\mathbb{Z}}$ consists of all pairs $((i, m), (j, n))$ such that one of the following holds:

(1) $i = 0 \wedge j = 1$;
(2) $i = 1 \wedge j = 1 \wedge L(m, n)$;
(3) $i = 0 \wedge j = 0 \wedge L(n, m)$.

Condition (1) says that all negative numbers $(0, m)$ are smaller than all nonnegative numbers $(1, n)$. Condition (2) says that the relation $\mathsf{Less}_{\mathbb{Z}}$ agrees with the relation

[2] $\mathbb{Z}$ for the German word Zahlen.

Less on the set of nonnegative numbers. Condition (3) says that the ordering of the negative numbers reverses the ordering of the positive ones. For example, $(0, 5)$ is less than $(0, 2)$, and this is what we want, because -5 is less than -2. Notice that all three conditions can be put together with the connective $\vee$ combining them into a single first-order statement.

Why, you can ask, are we making things so complicated? Why don't we just say that the integers are the natural numbers and their opposites? In other words, why don't we say that $\mathbb{Z}$ is the *union* of two sets, one being $\mathbb{N}$ itself, and the other defined as $\{-m : m \in \mathbb{N}\}$. We could do that, but then our task would be to explain the meaning of $-m$. What is $-m$? What is this operation of putting the negative sign in front of a natural number? Putting a negative *sign* in front of an expression is a syntactic operation, i.e. it is an operation on symbols. We are not defining sets of symbols, we are defining honest to goodness sets of mathematical objects. It is all related to the separation of syntax and semantics. We need to keep a clear dividing line between syntactic symbols and mathematical objects they represent. But is it really an honest separation? Instead of $-m$ we wrote $(0, m)$, and we said that $(0, m)$ is an *ordered pair*. Why is $(0, m)$ better than $-m$? That is explained by a convenient feature the syntax of first-order logic: it allows us to express facts about ordered pairs, triples, and finite tuples in general in a natural way. If x and y are free (unquantified) variables in a formula $\varphi(x, y)$, then, in any structure in which the language is interpreted, $\varphi(x, y)$ defines the set of ordered pairs having the property expressed by the formula. Hence, to define the structure $(\mathbb{Z}, \mathsf{Less}_{\mathbb{Z}})$ we did not need to expand our formal language; what we already have suffices.

Now, in a similar fashion we will define addition $\mathsf{Add}_{\mathbb{Z}}$, and multiplication $\mathsf{Mult}_{\mathbb{Z}}$ as relations on $\mathbb{Z}$. (If you are not intrigued, you can skip the details.)

The definition of $\mathsf{Add}_{\mathbb{Z}}$ given below refers only to addition and the order relation on the set $\mathbb{N}$; in other words, $\mathsf{Add}_{\mathbb{Z}}$ is first-order definable in the structure $(\mathbb{N}, \mathsf{Less}, \mathsf{Add})$. Since, the ordering relation Less on $\mathbb{N}$ is first-order definable in the structure $(\mathbb{N}, \mathsf{Add})$, it follows that $\mathsf{Add}_{\mathbb{Z}}$ is also first-order definable in $(\mathbb{N}, \mathsf{Add})$, i.e. the reference to Less in the definition below can be eliminated. In fact, this way we could eliminate all references to Less, and hence all occurrences of the relation symbol L as well, but if we replaced all occurrences of statements of the form $L(x, y)$ by their definition: $\exists z[z\neg(z = 0) \wedge A(x, z, y)]$, that would make the already complicated formulas almost impossible to read.

We will now formally define addition of integers. Addition of signed numbers is usually explained by examples, including all different cases. This is exactly what Definition 4.3 below does. For given three integer numbers in our chosen format, (h, k), (i, l), and (j, m), the definition list all cases in which (j, m) is the sum of (h, k) and (i, l). For example, if $k = 7$ and $l = 3$, the cases we need to include are: $7 + 4 = 10$, $(-7) + (-3) = (-10)$, $(-3) + 7 = 4$, $3 + (-7) = (-4)$. Since we usually tacitly assume that addition of integers is commutative, i.e. the result does not depend on the order in which numbers are added, once we are told that $(-3) + 7 = 4$, we also know that $7 + (-3) = 4$. In a formal definition all tacit assumptions have to be made explicit. Nothing can be left our. The result is a definition that looks very technical. If you are not intrigued by all those details, skip it, and instead move on to Definition 4.4.

Definition 4.3 The relation $\mathsf{Add}_{\mathbb{Z}}$ is the collection of all ordered triples $((h,k),(i,l),(j,m))$ such that one of the following holds,

(1) $h=1 \wedge i=1 \wedge j=1 \wedge A(k,l,m)$;
(2) $h=0 \wedge i=0 \wedge j=0 \wedge A(k,l,m)$;
(3) $h=1 \wedge i=0 \wedge k=l \wedge j=1 \wedge m=0$;
(3′) $h=0 \wedge i=1 \wedge k=l \wedge j=1 \wedge m=0$;
(4) $h=1 \wedge i=0 \wedge L(k,l) \wedge j=0 \wedge A(m,k,l)$;
(4′) $h=0 \wedge i=1 \wedge L(l,k) \wedge j=0 \wedge A(m,l.k)$;
(5) $h=1 \wedge i=0 \wedge L(l,k) \wedge j=1 \wedge A(m,l,k)$;
(5′) $h=0 \wedge i=1 \wedge L(k,l) \wedge j=1 \wedge A(m,k,l)$.

For completeness of the presentation, let us also define the multiplication relation $\mathsf{Mult}_{\mathbb{Z}}$. The definition is mercifully simpler than that of $\mathsf{Add}_{\mathbb{Z}}$.

Definition 4.4 The relation $\mathsf{Mult}_{\mathbb{Z}}$ is the collection of all ordered triples $((h,k),(i,l),(j,m))$ such that one of the following holds,

(1) $h=1 \wedge i=1 \wedge j=1 \wedge M(k,l,m)$;
(2) $h=0 \wedge i=0 \wedge j=1 \wedge M(k,l,m)$;
(3) $h=1 \wedge i=0 \wedge j=0 \wedge M(k,l,m)$;
(4) $h=0 \wedge i=1 \wedge j=0 \wedge M(k,l,m)$.

Notice that (2) expresses the rule "$negative \cdot negative = positive$."

For us, the most important feature of Definitions 4.3 and 4.4 is that the new structure, $(\mathbb{Z}, \mathsf{Add}_Z, \mathsf{Mult}_{\mathbb{Z}})$, has been defined completely in terms of the structure $(\mathbb{N}, \mathsf{Add}, \mathsf{Mult})$, or using our metaphor, $(\mathbb{Z}, \mathsf{Add}_Z, \mathsf{Mult}_{\mathbb{Z}})$ can be logically seen in the structure $(\mathbb{N}, \mathsf{Add}, \mathsf{Mult})$. We will see later that these two structures are even more intimately connected. Namely, not only is $(\mathbb{Z}, \mathsf{Add}_Z, \mathsf{Mult}_{\mathbb{Z}})$ first-order definable in $(\mathbb{N}, \mathsf{Add}, \mathsf{Mult})$, but also $(\mathbb{N}, \mathsf{Add}, \mathsf{Mult})$ is first-order definable in $(\mathbb{Z}, \mathsf{Add}_Z, \mathsf{Mult}_{\mathbb{Z}})$. We need to develop more tools to show how it is done.

4.3.1 Natural Numbers Are Integers, Naturally

According to our definition, integers are ordered pairs of pairs of natural numbers, so, as defined, the domains $\mathbb{N}$ and $\mathbb{Z}$ are disjoint. Usually, the natural numbers are identified with the non-negative integers. The structure with domain $\{(1,m) : m \in \mathbb{N}\}$ and the relations Add_Z and, $\mathsf{Mult}_{\mathbb{Z}}$ restricted to this domain, is a copy of the structure $(\mathbb{N}, \mathsf{Add}, \mathsf{Mult})$; the only difference is that every natural number n is now represented by the pair $(1,n)$. Later, we will be saying, more precisely, that we identify these two structures *up to isomorphism*. In this sense, the set of natural numbers $\mathbb{N}$ becomes a *subset* of the set of integers $\mathbb{Z}$.

What does it mean to identify to structures? In a precise way, the previous paragraph can be rephrased as follows.[3] The function $f : \mathbb{N} \longrightarrow \mathbb{Z}$, defined by $f(n) = (1, n)$ is a one-to-one correspondence between sets $\mathbb{N}$ and $\mathbb{Z}$, and for all $k, l, m \in \mathbb{N}$, $(k, l, m) \in \mathsf{Add}$ if and only if $(f(k), f(l), f(m)) \in \mathsf{Add}_{\mathbb{Z}}$, and $(k, l, m) \in \mathsf{Mult}$ if and only if $(f(k), f(l), f(m)) \in \mathsf{Mult}_{\mathbb{Z}}$. This shows that the image of $\mathbb{N}$ under f is *isomorphic* to $(\mathbb{N}, \mathsf{Add}, \mathsf{Mult})$.

The ordering relation Less is defined in $(\mathbb{N}, \mathsf{Add})$ by the formula $\exists z[z\neg(z = 0) \wedge A(x, z, y)]$. This formula does not define the $\mathsf{Less}_{\mathbb{Z}}$ in $(\mathbb{Z}, \mathsf{Add}_{\mathbb{Z}})$. In $(\mathbb{Z}, \mathsf{Add}_{\mathbb{Z}})$ it defines the set of all integers. In fact, it can be shown that Less_Z can not be defined in $(\mathbb{Z}, \mathsf{Add}_{\mathbb{Z}})$ by any first-order formula at all. This interesting result is not difficult to prove, and we will do it later after we discuss a general strategy for proving such results.

We are building numbers systems, from the natural numbers up. So far, we made one step, from the natural numbers to the integers. There will be more steps, each justified by a particular need in the historical development of mathematics. How the number systems came to be what they are now is a fascinating story, and we will not discuss it fully. For a comprehensive account, including technical details that are omitted here, see [32].

4.4 Fractions!

While natural numbers have their origin in counting and multiplying, the rational numbers, originate from cutting, slicing, and breaking. If you break a chocolate bar into eight equal pieces, and eat three of them, you will have eaten is $\frac{3}{8}$ of the whole bar. If you eat one more, this will add up to $\frac{3}{8} + \frac{1}{8} = \frac{4}{8}$ of the bar. The result is the same as if you broke the bar into two equal pieces, and ate one of them. The fraction $\frac{4}{8}$ represents the same quantity as $\frac{1}{2}$.

All basic properties of addition and multiplication of fractions can be explained on simple examples, as we did above. For practical applications, such explanations suffice to justify the rules for adding and multiplying fractions. It all seems simple and natural, but, as every teacher of mathematics knows, the arithmetic of fractions often instills fear and loathing in students, and this phenomenon is not that easy to explain. We will not get into this discussion here, but what follows is relevant. As you will see, a full description of the arithmetic structure of fractions, requires significantly more effort than what suffices for addition and multiplication of integers.

In the next section, we will perform an exercise in the formalism of first-order logic. We will define a structure of fractions inside $(\mathbb{Z}, \mathsf{Add}_Z, \mathsf{Mult}_{\mathbb{Z}})$, and therefore also in $(\mathbb{N}, \mathsf{Add}, \mathsf{Mult})$. This definition will be based on a useful feature of formal

---

[3] Skip this paragraph if the terms and notation are unfamiliar. This whole topic will be thoroughly discussed in Part II.

logic—its ability to deal with abstraction in a precise way. The problem we need to address is that fractions such as $\frac{6}{4}$ and $\frac{3}{2}$ are not equal as formal expressions. They are not the same, but they do represent the same quantity, so they are in this sense *equivalent*. There is great redundancy in fractions. The same quantity can be represented by infinitely many different but equivalent fractions. Because of that, we will make a distinction: by a fraction we will mean a formal expression of the form $\frac{m}{n}$, and by a *rational number* for now we will mean the quantity that a fraction represents. So we have to address another question: What is the quantity represented by a fraction? We will give a precise answer, but the reader should be warned again that this material is technically more advanced. If you have never liked fractions, you should skip the next section, and return to it later.

4.5 The Arithmetic Structure of the Rationals

We will define the structure $(\mathbb{Q}, \mathsf{Add}_{\mathbb{Q}}, \mathsf{Mult}_{\mathbb{Q}})$, whose domain is the set of rational numbers and $\mathsf{Add}_{\mathbb{Q}}$ and $\mathsf{Mult}_{\mathbb{Q}}$ are the addition and multiplication as relations on $\mathbb{Q}$. All definitions will take place in $(\mathbb{Z}, \mathsf{Add}_{\mathbb{Z}}, \mathsf{Mult}_{\mathbb{Z}})$.

Let $\mathcal{F}$ be the set of all ordered pairs of integers (m, n), where n is not 0. The intention here is that (m, n) represents the fraction $\frac{m}{n}$. The domain of the rational numbers will be defined shortly, for now we will describe the additive and multiplicative structure on the set $\mathcal{F}$.

Recall that adding fractions requires a procedure. The sum of $\frac{k}{l}$ and $\frac{m}{l}$ is $\frac{k+m}{l}$, but if the denominators are different, we first need to find equivalent fractions with a common denominator, and then add their numerators. This whole process is summarized in the following addition formula:

$$\frac{k}{l} + \frac{m}{n} = \frac{k \cdot n + l \cdot m}{l \cdot n}.$$

For example $\frac{1}{2} + \frac{1}{3} = \frac{1 \cdot 3 + 1 \cdot 2}{2 \cdot 3} = \frac{5}{6}$.

Translated into the language of pairs and relations the rule for multiplication gives the following definition of the addition relation for $\mathcal{F}$: $((k, l), (m, n), (r, s)) \in \mathsf{Add}_{\mathcal{F}}$ whenever

(1) $k \cdot n + l \cdot m = r$ and
(2) $l \cdot n = s$.

In the two conditions above, $+$ and $\cdot$ are the usual addition and multiplication. They are functions, and since we do not allow function symbols in formal statements, we have to pay a price now, when we translate the above definition into the language with relation symbols A and M. The translated formula is not pretty. The addition relation $\mathsf{Add}_{\mathcal{F}}$ can be defined as the set of triples of pairs of integers $((k, l), (m, n), (r, s))$ such that for some a and b in $\mathbb{Z}$, $k \cdot n = a$, $l \cdot m = b$, $a + b = r$, and $l \cdot n = s$. Written in the language of relations, the first-order definition of $\mathsf{Add}_{\mathcal{F}}$

over $(\mathbb{Z}, \mathsf{Add}_{\mathbb{Z}}, \mathsf{Mult}_{\mathbb{Z}})$ is given by the formula

$$\exists a \exists b\, [M(k, n, a) \wedge M(l, m, b) \wedge A(a, b, r) \wedge M(l, n, s)].$$

We have defined the addition relation on the set $\mathcal{F}$ in terms of addition and multiplication on $\mathbb{Z}$.

The complications above are caused by the fact that the algorithm for addition of fractions requires finding their common denominator. The rule for multiplication of fractions is more user friendly. To define the multiplication relation $\mathsf{Mult}_{\mathcal{F}}$ we can use the familiar rule:

$$\frac{k}{l} \cdot \frac{m}{n} = \frac{k \cdot m}{l \cdot n}.$$

This shows that multiplication in $\mathcal{F}$ is fully determined by multiplication in $\mathbb{Z}$, and it is a routine exercise to write a first-order formula that defines $\mathsf{Mult}_{\mathcal{F}}$ in $(\mathbb{Z}, \mathsf{Mult}_{\mathbb{Z}})$.

Finally, let us define the ordering of fractions. Do you remember how to find out which of two given fractions is larger? For example, which one is larger, $\frac{7}{13}$ or $\frac{5}{11}$? The problem here is that the first fraction represents thirteenths of a unit, and the second, elevenths. They are like apples and oranges, you cannot compare them directly. To compare the fractions, we need convert them to equivalent fractions with a least common denominator, which in this case is $13 \cdot 11 = 143$. Lets do it: $\frac{7}{13} = \frac{7 \cdot 11}{13 \cdot 11} = \frac{77}{143}$; and $\frac{5}{11} = \frac{5 \cdot 13}{11 \cdot 13} = \frac{65}{143}$, and we can see that $\frac{7}{13}$ is larger.

The method described above gives us the following criterion for comparing fractions. If k, l, m, and n are positive integers, then $\frac{k}{l} < \frac{m}{n}$ if and only if $k \cdot n < l \cdot m$. This shows that the ordering of positive fractions can be defined in $(\mathbb{Z}, \mathsf{Less}_{\mathbb{Z}}, \mathsf{Mult}_{\mathbb{Z}})$. Then the definition can be extended to all fractions by considering cases, keeping in mind that if both m and n are negative, then $\frac{m}{n}$ is positive, and if only one of them is negative, then $\frac{m}{n}$ is negative as well.

4.5.1 Equivalence Relations and the Rationals

Abstraction is a process in which new concepts are arrived at by stripping the objects of discourse of some of their features. Mathematical logic has a formalized approach to this kind of abstraction and this approach is essential to many developments in modern mathematics.

In abstracting, we identify object that are considered equivalent according to some criteria. The following definition makes this precise.

Definition 4.5 A set E of ordered pairs of elements of a set A is an *equivalence relation* on A if it satisfies the following three conditions:

(1) E is *reflexive* i.e. for all $a \in A$, $(a, a) \in E$, i.e. every element is related to itself;

(2) E is *symmetric* i.e. for all $a, b \in A$, if $(a, b) \in E$, then $(b, a) \in E$, i.e. if a is related to b, then b is related to a;
(3) E is *transitive* i.e. for all $a, b, c \in A$, if $(a, b) \in E$ and $(b, c) \in E$, then $(a, c) \in E$, i.e. if a is related to b, and b is related to c, then a is related to c.

If E is an equivalence relation and (a, b) is in E, then we say that a and b are *E-equivalent*.

To discuss properties of equivalence relations, we need to introduce some notation. The next two paragraphs should be read as an exercise. Try to see how the facts stated below follow directly from Definition 4.5.

Let E be an equivalence relation on a set A. Then for each $a \in A$, the set $\{b : b \in A \wedge (a, b) \in E\}$ is called the *equivalence class* of a, and is denoted by $[a]_E$. For each equivalence class any of its elements is a *representative* of the class; in particular, a is always a representative of $[a]_E$. If a and b are E-equivalent, then $[a]_E = [b]_E$. Moreover, for all a and b in A, either the equivalence classes $[a]_E$, $[b]_E$ coincide, or they are disjoint. It follows that the set of all equivalence classes forms a partition of the set A into disjoint nonempty subsets.

For another exercise, you can go over Definition 4.5 and convince yourself that the relation of being related (in the family sense) is an equivalence relation on the domain of all people. Notice that one has to make a somewhat controversial assumption that everyone is related to herself or himself. What are the equivalence classes of this relation? What is your equivalence class? How many representatives does it have? This last question has a theoretical answer, but the exact answer is impossible to pin down in practice.

Now let us define the following relation Eq on the set of fractions $\mathcal{F}$

$$((k, l), (m, n)) \in \text{Eq} \text{ if and only if } k \cdot n = l \cdot m.$$

One can show, and the reader is encouraged to do it as an exercise, that Eq is an equivalence relation. Observe that transitivity property (3) holds because pairs of the form $(m, 0)$ are not fractions. See what goes wrong if they were.

Since Eq is an equivalence relation, we can talk about its equivalence classes. For example $[(1, 2)]_{\text{Eq}} = \{(1, 2), (-1, -2), (2, 4), (-2, -4), \dots\}$.

So far we talked about fractions, now it is time to define the rational numbers.

Definition 4.6 *Rational numbers* are equivalence classes of the relation Eq on the set of fractions $\mathcal{F}$. In the set notation, the set of rational numbers $\mathbb{Q}$ is

$$\{[(m, n)]_{\text{Eq}} : (m, n) \in \mathcal{F}\}.$$

In mathematics, when we say $\frac{1}{2}$, we often mean the one-half as a rational number, as in our example $[(1, 2)]_{\text{Eq}}$ above. In other words, we think of it not just as a single fraction, but also as the whole set of fractions that are equivalent to it. This practice is formalized with the help of the equivalence relation Eq.

4.5.2 *Defining Addition and Multiplication of the Rational Numbers Formally*

Addition and multiplication of the integers determines addition and multiplication on the set of fractions $\mathcal{F}$. Because of the more complex nature of the rational numbers, defining their arithmetic operations is harder. This section shows how it is done. This is the most technical section of this chapter and it can be skipped on the first reading.

We define addition and multiplication of rational numbers as ternary relations as follows:

$$\mathsf{Add}_{\mathbb{Q}} = \{([(h,k)]_{\mathrm{Eq}}, [(i,l)]_{Eq}, [(j,m)]_{\mathrm{Eq}}) : ((h,k),(i,l),(j,m)) \in \mathsf{Add}_{\mathcal{F}}\},$$

$$\mathsf{Mult}_{\mathbb{Q}} = \{([(h,k)]_{\mathrm{Eq}}, [(i,l)]_{\mathrm{Eq}}, [(j,m)]_{\mathrm{Eq}}) : ((h,k),(i,l),(j,m)) \in \mathsf{Mult}_{\mathcal{F}}\}.$$

Notation becomes really heavy here, and that is because we are talking about a domain whose elements are equivalence classes of elements from another domain. Those two relations on $\mathbb{Q}$ are defined in terms of the already defined addition and multiplication on the set of fractions $\mathcal{F}$. It is not immediately clear that the definitions are correct. We defined sets of ordered triples of equivalence classes of fractions, but in the conditions defining the relations instead of properties of classes, we use properties of their representatives. A proof is needed that those conditions do not depend on the choice of representatives. This is one of the exercises at the end of this chapter.

Similarly, the ordering of $\mathbb{Q}$ is determined by the ordering of $\mathcal{F}$:

$$\mathsf{Less}_{\mathbb{Q}} = \{([(k,l)]_{\mathrm{Eq}}, [(m,n)]_{\mathrm{Eq}}) : ((k,l),(m,n)) \in \mathsf{Less}_{\mathcal{F}}\}.$$

As above, one can check that the definition does not depend on the choice of representatives of equivalence classes.

In elementary mathematics such a level of pedantry as we exercised above is not necessary, and it could be harmful. It is worth noting though that some specialists in mathematics education maintain that one of the reasons that the arithmetic of rational numbers is a difficult topic is that the definition of $\mathbb{Q}$ involves equivalence classes, rather than just individual fractions.

The move from a set with an equivalence relation on it to the set of equivalence classes, called "taking the quotient," is an important mathematical operation. It is a mathematical formalization of the process of passing from the particular to the general.

As formally defined, the sets $\mathbb{N}$ and $\mathbb{Z}$ are disjoint, but, as we saw, there is a way in which one can identify natural numbers in $\mathbb{N}$ with the positive numbers in $\mathbb{Z}$; hence we can consider $\mathbb{Z}$ as an extension of $\mathbb{N}$. Similarly, by identifying each integer m with the rational number $[(m,1)]_{\mathrm{Eq}}$, we consider $\mathbb{Q}$ to be an extension of $\mathbb{Z}$.

Notice that in the definition of $\mathbb{Q}$ at the end of the previous section, we moved away from the structure of the integers. We changed the domain quite significantly. It is defined as a set of equivalence classes of pairs of integers, and those classes are infinite sets. Every fraction has infinitely many fractions that are equivalent to it, and hence, each equivalence class $[(m, n)]_{\mathrm{Eq}}$ is infinite. First-order definitions can be used to define sets of elements, ordered pairs, ordered triples, and, in general, ordered sequences of any fixed finite length, but not sets of infinite sets. Fractions, the elements of $\mathcal{F}$ are just certain pairs of integers, but the domain $\mathbb{Q}$ is the set of infinite sets. Still, there is a way to ground the new structure in the old one in a first-order fashion, by selecting a representative for each equivalence class in $\mathbb{Q}$ in a certain way.

A fraction $\frac{m}{n}$ is *reduced* if m and n do not have common factors. For example, $\frac{2}{5}$ is reduced, but $\frac{6}{15}$ is not, because 3 is a common factor of 6 and 15. Each equivalence class $[(m, n)]_{\mathrm{Eq}}$ has only one reduced fraction in it, and it can be chosen to be its representative. Moreover, the set of reduced fractions is first-order definable in $(\mathbb{Z}, \mathsf{Add}_Z, \mathsf{Mult}_{\mathbb{Z}})$ (this is left as an exercise). Addition and multiplication of reduced fractions matches addition and multiplication of the equivalence classes they represent, so we obtain a copy of the structure of the rational numbers that is logically visible in $(\mathbb{Z}, \mathsf{Add}_{\mathbb{Z}}, \mathsf{Mult}_{\mathbb{Z}})$.

4.5.3 Dense Ordering of the Rationals

The ordering of the natural numbers is easy do describe. The natural numbers begin with zero, and progress up in increments of one. The ordering of the integers is similar, they also can be thought of as beginning with zero, and then going in two directions up and down (or left and right) in increments of one. The key property here is that every natural number and every integer m has a unique successor $m + 1$. This property makes $(\mathbb{N}, \mathsf{Less})$ and $(\mathbb{Z}, \mathsf{Less}_{\mathbb{Z}})$ discrete orderings.

The ordering of $\mathbb{Q}$ is different. Between any two rational numbers there is another one. For example, if p and q are rational and $p < q$, then $\frac{p+q}{2}$ is a rational number between p and q.

In general, if a relation R linearly orders a set M, the ordering R is called *dense* if between any two elements of M there is another element. Formally, R is dense if for all a and b in M, if $(a, b) \in \mathsf{R}$, then there is a c such that $(a, c) \in \mathsf{R}$ and $(c, b) \in \mathsf{R}$.

Density of the ordering of $\mathbb{Q}$ has important consequences, and they will be explored in the next chapter. For now, let us just notice that density implies that not only is the whole set $\mathbb{Q}$ actually infinite, but also that already between any two rational numbers there are infinitely many rational numbers. Because of that, illustrating the ordering of rational numbers as points on a number line is not easy. On the one hand, there should not be any visible gaps in the line, as between any two points there have to be (infinitely many!) other points. On the other hand, as you will see in the next chapter, the line made of points corresponding to rational numbers cannot be solid. For reasons having to do with geometry it must have gaps. This is a serious problem.

Exercises

Exercise 4.1 *Show that if an element in a linearly ordered set has an immediate successor, than it has only one immediate successor.*

Exercise 4.2 *Let (M, R) be a linearly ordered set. Assume that the statement $\forall m \exists n\, L(m, n)$ holds in (M, R). Prove that the set M is infinite.*

Exercise 4.3 *The argument below is a sequence of true statements about* 0, 1, *and* -1. *Provide a justification for each step.*

$$1 + (-1) = 0,$$
$$(-1) \cdot (1 + (-1)) = 0,$$
$$(-1) \cdot 1 + (-1) \cdot (-1) = 0,$$
$$(-1) + (-1) \cdot (-1) = 0,$$
$$1 + (-1) + (-1) \cdot (-1) = 0 + 1,$$
$$(-1) \cdot (-1) = 1.$$

Exercise 4.4 *According to the definition of* $\mathbb{Z}$ *in this chapter, the integers* 2, 5, -2, *and* -5 *are represented by* $(1, 2)$, $(1, 5)$, $(0, 2)$, *and* $(0, 5)$, *respectively. Use this representation and Definition* 4.3 *to compute* $(-2) + (-5)$, $(2) + (-5)$, *and* $(-2) + (5)$.

Exercise 4.5 * *Verify that Definition* 4.3 *is correct.*

Exercise 4.6 *Define* $\mathsf{Mult}_{\mathbb{Q}}$ *in terms of* $\mathsf{Mult}_{\mathbb{Z}}$.

Exercise 4.7 *Prove that if E is an equivalence relation on a set A, then for all a and b in A, either the equivalence classes* $[a]_E$ *and* $[b]_E$ *coincide, or they are disjoint.*

Exercise 4.8 *Show that the relation E on the set of fractions* $\mathcal{F}$ *defined by:* $(\frac{k}{l}, \frac{m}{n}) \in E$ *if and only if* $k \cdot n = l \cdot m$, *is an equivalence relation.*

Exercise 4.9 *Define the ordering* $\mathsf{Less}_{\mathbb{Q}}$ *in terms of* $\mathsf{Less}_{\mathbb{Z}}$ *and* $\mathsf{Mult}_{\mathbb{Z}}$.

Exercise 4.10 * *Find a first-order formula defining the set of reduced fractions in* $(\mathbb{Z}, \mathsf{Add}_Z, \mathsf{Mult}_{\mathbb{Z}})$.

Exercise 4.11 *This exercise requires some elementary algebra. Use formal definitions of* $\mathsf{Less}_{\mathbb{Q}}$, $\mathsf{Add}_{\mathbb{Q}}$ *and* $\mathsf{Mult}_{\mathbb{Q}}$ *to verify that for all distinct rational numbers p and q,* $\frac{1}{2} \cdot (p + q)$ *is between p and q.*

Chapter 5
Points, Lines, and the Structure of $\mathbb{R}$

Abstract In the previous chapter we saw how a large portion of mathematics can be formalized in first-order logic. The very fact that the construction of the classical number structures can be formalized this way makes first-order logic relevant, but is it necessary? For centuries mathematics has been developing successfully without much attention paid to formal rigor, and it is still practiced this way. When intuitions don't fail us, there is no need for excessive formalism, but what happens when they do? In modern mathematics intuition can be misleading, especially when actual infinity is involved. In this chapter, we will see how seemingly innocuous assumptions about actually infinite sets lead to consequences that are not easy to accept. Then, we will go back to our discussion of a formal approach that will help to make some sense out of it.

Keywords Square root of 2 · Irrational numbers · Real numbers · Dedekind cuts · Dedekind complete orderings · Banach-Tarski paradox · Infinite decimals

5.1 Density of Rational Numbers

The rational numbers are ordered densely. Between any two rational numbers, there is another one. To illustrate this graphically, we can start marking points on a straight line, starting with two, and continuing by marking new points in-between points already marked. Soon, no matter how fine our marks are, the line begins to look solid. If we imagine that all rational numbers have been marked, it seems that there should be no space left for any other marks. How reliable is that intuition? We should be careful. When we think of a point, we see it as a dot on a plane. We see lines with different widths, some can be thinner, some thicker. Such images are helpful visualizations of ideal geometric objects, but they are not the objects themselves. A geometric point cannot be seen. It has no length nor width. It is just an idea of a perfect, exact location. Similarly, lines have no width, they only have length. We will examine this more carefully in a moment, but now it is time for a short digression on sets that are actually infinite.

R. Kossak, *Mathematical Logic*, Springer Graduate Texts in Philosophy 3,
https://doi.org/10.1007/978-3-319-97298-5_5

The sets $\mathbb{N}$, $\mathbb{Z}$, and $\mathbb{Q}$ share a common feature: they can be generated in an infinite step-by-step process. What is an infinite step-by-step process? We have already seen one. This is how the natural numbers are made: in the first step, we construct the number zero; in the second step the number one, and son on. In each next step we construct a number one larger than the previous one. This way, sooner or later, each natural number gets constructed. Now imagine that all those steps have been performed. This can be thought of as a magic limit step ω. In the first step, the first operation is performed, in the second—second, and so on. Then there is the step ω, marking the fact that all possible finite steps have been performed. It is the first infinite step. Here we speak about it metaphorically, later it will become a legitimate set-theoretic notion.

I have not explained what kind of constructions are allowed in step-by-step constructions, or even what it means to "construct a number," this will be made clearer in Chap. 6. Mathematical constructions are mental, although they are often modeled on actual operations, such as counting, performing a geometric construction, or calculating according to some formula. For now let us just think of any mental process in which steps are clearly understood, and that does not involve anything actually infinite.

Here is how we can construct $\mathbb{Z}$: in step one, we construct $\{0\}$ in step two $\{-1, 0, 1\}$, in step three, $\{-2, -1, 0, 1, 2\}$, ..., in step 100 we get the set $\{-99, -98, -97, \dots, 97, 98, 99\}$, and so on. All these sets are finite approximations to our goal. We will have "constructed" all of $\mathbb{Z}$ in the limit step ω.

The process of generating the set of all fractions $\mathcal{F}$ step-by-step is a little more complicated. There are many ways to do it. Here is one.

- In step one, we make $\{0\}$.
- In step two, $\{-\frac{1}{1}, 0, \frac{1}{1}\}$.
- In step three, we make all fractions that can be written with numerators and denominators less than three: $\{-\frac{2}{2}, -\frac{1}{1}, -\frac{1}{2}, 0, \frac{1}{2}, \frac{1}{1}, \frac{2}{1}\}$.
- In step four we make the set of all fractions whose numerators and denominators are less than four.
- and so on...

In each step, we systematically add more and more fractions, with larger and larger numerators and denominators—finitely many at each step—making sure that every fraction will appear sooner or later. This is a well-defined process that will eventually generate all fractions, including all those different fractions that are equivalent to each other. In the limit step ω we will obtain all fractions. If we want $\mathbb{Q}$ as the final result, each step can be followed by erasing redundancies such as $\frac{5}{5}$ or $-\frac{2}{4}$.

A geometric line, or a line segment, is a single geometric object, but in modern mathematics we think of it as made of points, so it becomes an actual (not just potential) infinite set. How many points make a line? Certainly infinitely many, but we will try to be more precise. We can think of straight line as an ideal measuring tape. One point is designated to represent 0, then there are infinitely many points spread at equal distances in both directions away from 0. Those represent all integers; positive going in one direction, negative in the other. Usually we imagine this line positioned horizontally, with the positive numbers to the right, and negative to the left. We can mimic the step-by-step construction of the rational numbers and consecutively mark all rational numbers, with each number marked according to the magnitude it represents. For example $\frac{1}{2}$ is marked in the middle between 0 and 1, and the place for $\frac{4}{3}$ is found by dividing the segment between 1 and 2 into three equal pieces and marking $\frac{4}{3}$ at the end of the first piece to the right of 1. This is a laborious process, but it is easy to imagine how it all can be done by geometric means in a step-by-step fashion. It is interesting that the whole construction can be done just with a ruler and a compass (and infinite time on your hands). It is not immediately obvious how to divide a line segment into, for example, 13 equal pieces just with a ruler and a compass, but it can be neatly done. You can see it in animation at http://www.mathopenref.com/constdividesegment.html.

We will say that a point of the line is *rational* if it is one of the points marked in the process described above. The rational points are geometric representations of all rational numbers. The left-to-right ordering of the rational points agrees with the larger-smaller ordering of the rational numbers they represent. It is a common mathematical practice to identify the rational numbers with their geometric representations.

Is every point on the geometric line a rational point? This question would be hard to answer if we relied only on very basic geometric intuitions. We need a deeper insight. Consider a square with each side of length 1 positioned so that its lower left corner coincides with the point marked 0 on the number line (see Fig. 5.1). The circle whose center is at 0 and which passes through the upper right vertex of the square crosses the number line at a point. It was one of the great discoveries of Pythagorean mathematics (attributed to a member of the Pythagorean school named Hippasus) that point of intersection of the circle and the line is not rational. It follows that the rational numbers do not cover the whole line. The Pythagoreans did not think about it this way; what they proved was that the diagonal of a square is *incommensurable* with its side.[1] We will talk about incommensurability in the next section.

So now other questions suggest themselves. If there are points on the line that are not rational, how many such points are there? Moreover, since the line is not made of rational points, what else is it made of? If you think of the non-rational points as gaps in the line that is made only of the rational points, can those gaps be filled to complete the line? What does it mean to fill a gap? If a solid geometric line is a *continuum* without gaps, can it be made of points? If it is made of points, how is it made? If there are many points on the geometric line that do not correspond to rational numbers, do they correspond to some other numbers? What numbers? Where do those other numbers come from?

[1] This geometric fact has an interesting number theoretic variant: the sum of two equal square numbers is never a square number. $1^2 + 1^2 = 1 + 1 = 2$ (not a square), $2^2 + 2^2 = 4 + 4 = 8$ (not a square), $3^2 + 3^2 = 9 + 9 = 18$ (not a square), and so on. An elegant proof of this is in Appendix A.2.

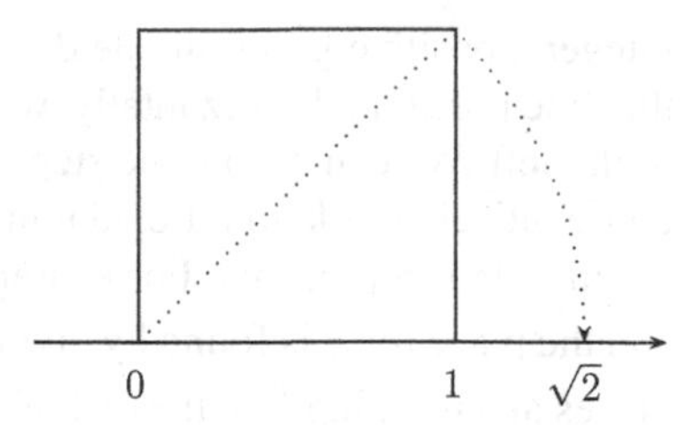

Fig. 5.1 $\sqrt{2}$ finds its place on the number line

5.2 What Are Real Numbers, Really?

Numbers which correspond to "gaps" in the line made of rational points are called *irrational*. All numbers, rational and irrational, that can be represented as distances between points on the real line, are called *real numbers*. The real line here is what in the previous section we called a geometric continuous line. This rather vague definition of the real numbers is usually given in entry level mathematics courses, and it suffices for many practical purposes. Actually, the intuition behind it suffices for much more, including some advanced mathematics. However, here we are concerned with a more careful approach. Notice that if we declare, as it is routinely done, that real numbers are distances from the origin (the point marking zero) to other points on the geometric line, then we are obliged to explain what we mean by a distance. Those distances are numbers, one usually hears, but this kind of answer will certainly not satisfy us.

The Pythagoreans discovered that the length of the diagonal of the unit square, is not commensurable with the base of the square. To explain what it means, let us consider the square in Fig. 5.1. What is the length of the diagonal of the square? By inspecting the picture, we can see that its length should be between 1 and 2, so its measure will not be a whole number. Will it be a fraction then? If the base of the square is divided into 10 equal intervals, one can check that 14, of such intervals, can be placed one by one on the diagonal, but the 15th sticks out. If the picture is large and precise enough, one can also see that the 14 intervals do not quite cover the diagonal. This means that the length of the diagonal is between 1.4 and 1.5. We could try a finer measure, say hundredths. When the base is divided into 100 intervals of equal length, one can verify that the diagonal can be almost covered with 141 such intervals, but the 142nd sticks out. This shows that the length of the diagonal is between 1.41 and 1.42. The meaning of incommensurability of the diagonal is that no mater what fraction of the base of the square we take as the unit measure, we can never find a number of intervals of this size that will exactly cover the diagonal. The last piece will either not reach the opposite corner, or will stick out. Surely though, everyone would agree that there should be a number that represents the length of the diagonal of the unit square. Let us call this number d.

It follows from the Pythagorean theorem that $1^2 + 1^2 = d^2$. Hence $d^2 = 2$, and because of that we will say that $d = \sqrt{2}$, which means that d is that number whose square equals 2. So let us be clear. At his point we know that $\sqrt{2}$ represents the length of the diagonal. Calling it a *number* is a bit premature, since we do not know what kind of number it is, and since it is not rational, it is not clear at all what it means to square it. We will deal with this problem in a moment, but for now notice, that the task of assigning a number value to the length of the diagonal of the unit square forces us to extend the number system beyond the rational numbers. We need at least one new number d, but that forces us to add more numbers. How many? At least infinitely many. Here is why. Let d_n be the length of the diagonal of an n by n square, where n is a natural number. From the incommensurability of the diagonal of a square with its side, it follows that d_n is not a rational number. Geometry forces us to include all those new numbers in a larger number system, but what are those numbers and how can they be included?

Let us go back to the idea that all points on the continuous geometric line correspond to numbers. Georg Cantor—the inventor of set theory—proved in a paper published in 1874 that there are more points on the line than there are rational numbers. What this means can be explained as follows. As we have seen, the set of all rational numbers can be constructed in a step-by-step process. The particular choice of what we do in each step is not important; what matters is that each step involves only finitely many operations. Cantor proved that the real line cannot be built from points in a step-by-step construction.[2] There are far too many points on the line which do not correspond to rational numbers. Cantor's argument is presented in Appendix A.2.

5.3 Dedekind Cuts

If we cannot construct a geometric line step-by-step, then how can we do it? The line must be rather complex, but can it be somehow constructed or built up from simpler pieces? Can it be made of points?

In modern mathematics, there are essentially two standard constructions of the real line. One proceeds via Dedekind cuts, and the other via equivalence classes of Cauchy sequences. Cauchy's construction is useful in mathematical analysis but it is more technical, so we will only describe what Dedekind did. In preparation, recall that in the previous chapter we defined rational numbers to be pairs of integers. Objects of a new kind—the rational numbers—were built is a certain way, from previously introduced objects—the integers. So, in this sense, in the hierarchy of mathematical objects, the rational numbers are more complex than the integers. Something similar will happen now. We will build the real numbers from the rational ones. The construction will be much less direct.

[2]An interesting technical aspect is that we might as well allow a step-by-step construction in which each step is itself a step-by-step construction. Cantor's theorem applies in this case as well.

Look again at Fig. 5.1. The number we called the square root of 2 is not rational, but its place on the number line can be determined by its position with respect to all rational numbers around it. Some are below it, some above. This leads to an idea to *identify* $\sqrt{2}$ with its place in the ordered set or rational numbers, and, further, to *identify* this place with a pair of sets. The set of rational numbers to the left of that place, and the set rational numbers to the right. Let us call the set to the left the Dedekind cut of $\sqrt{2}$, and let us denote it by $D_{\sqrt{2}}$.

Guided by geometry, we were lead to consider lengths of linear segments that cannot be measured using rational numbers. It seems natural to identify such lengths with places on the number line made of rational numbers, especially if such places can be determined by geometric means. Notice that even though originally $\sqrt{2}$ can be thought of a gap, as an empty space in the line made of rational numbers, this place is uniquely determined by $D_{\sqrt{2}}$; hence it is not nothing. The cut $D_{\sqrt{2}}$ is a well-defined mathematical object, and it is this object that we can also think of as $\sqrt{2}$. We can think of the set $D_{\sqrt{2}}$ as a number and we will make it more precise next.

We defined $\sqrt{2}$ using the construction illustrated on Fig. 5.1. The next goal is not only to define more such numbers, but to create a whole number system. It seems that to do that one would be forced to consider all situations in geometry, and in other areas of mathematics, where irrational numbers can occur, and to extend the number system by adding new numbers by considering more Dedekind cuts, or perhaps by some other means as well. To foresee what other irrational numbers might be needed would require exact definitions of ways in which such new numbers could be introduced. That would be very hard to do, but, to a great relief, it turns out unnecessary. Following Richard Dedekind, we can do something else. We will extend the system of rational numbers, by adding the largest possible set of Dedekind cuts. Before we do it in full generality, let us take a closer look at $D_{\sqrt{2}}$.

To define $D_{\sqrt{2}}$, one does not have to appeal to Fig. 5.1. Since $\sqrt{2}$ is not rational, for every rational number p either $p^2 < 2$ or $p^2 > 2$. We can define the Dedekind cut $D_{\sqrt{2}}$ as the set of rational number p such that $p^2 < 2$. The cut $D_{\sqrt{2}}$ defined this way does not have a largest element, and there is no smallest element above it. To see that this is the case, one must prove that for any rational p such that $p^2 < 2$, there is a rational q such that $p < q$ and $q^2 < 2$; and that any rational p such that $p^2 > 2$, there is a rational q such that $q < p$ and $2 < q^2$. Using elementary algebra, one can check that for $q = \dfrac{2p+2}{p+2}$, if $p^2 < 2$, then $p < q$, and $q^2 < 2$; and if $2 < p^2$, then $q < p$, and $2 < q^2$.

For a rational number p, let the Dedekind cut of p, denoted by D_p, be the set of all rational number that are less that p. For example, D_0 is the set of all rational negative numbers. Now we will consider a new structure whose elements are all sets D_p, and whose relations are defined in accordance with the ordering, addition, and multiplication in $\mathbb{Q}$. For example, for rational p, q, and r, we define that $D_p + D_q = D_r$ if and only if $p + q = r$. This way we obtain a copy of the $(\mathbb{Q}, \mathsf{Add}_{\mathbb{Q}}, \mathsf{Mult}_{\mathbb{Q}})$

in which each number p is replaced with the set of numbers D_p. The elements have changed, but they stand in a one-to-one correspondence with the old elements, and the correspondence $p \leftrightarrow D_p$ preserves the arithmetic relations. The new structure is an *isomorphic copy* of the old one. In this sense it is the same old $\mathbb{Q}$, with its addition and multiplication, and we will treat is as such. The point of this maneuver is to allow a smoother transition to the extension of $\mathbb{Q}$ to the larger set of real numbers.

For each rational p, there is a smallest rational number that is not in D_p. That number is p itself. The cut $D_{\sqrt{2}}$ is different. As we saw above, there is no smallest rational number p such that $2 < p^2$.

In general, a subset D of $\mathbb{Q}$ *Dedekind cut* if D is nonempty set, D is not the whole $\mathbb{Q}$, for every p, if p is in D then so are all rational numbers that are less than p, and D has no largest number.

Now we can say precisely what the real numbers are. They are exactly the Dedekind cuts. The real numbers are special sets of rational numbers. Some of them, those of the form D_p, for rational p, represent rational numbers, and all other, such that there is no smallest rational number above them, are *irrational*. We will use $\mathbb{R}$ as the name for the set of all Dedekind cuts. To make the set $\mathbb{R}$ a number structure, we need to say how the Dedekind cuts are added and multiplied. We will only define addition. This definition of multiplication is similar, but a bit more complicated. For details see [32].

If D and E are Dedekind cuts, then their sum $D+E$ is defined to be the Dedekind cut consisting of all sums $p+q$, where p is in D and q is in E. One has to check that the addition thus defined has all the required properties, such as $D + E = E + D$. It all works out, and as a result we have extended our number system to include all real numbers, and it is a huge accomplishment.

5.3.1 Dedekind Complete Orderings

If D and E are Dedekind cuts, then D is less that E if there is a p in E such that $q < p$, for all q in D. This gives us a new linearly ordered set $(\mathbb{R}, \mathsf{Less}_{\mathbb{R}})$. The ordering relation $\mathsf{Less}_{\mathbb{R}}$ is both dense and complete. Density of has already been defined. The notion of completeness can be formalized in different ways. We will present one.

In an attempt to keep the separation of syntax and semantics clearly visible, when comparing numbers, instead of the usual $x < y$, have been using $(x, y) \in \mathsf{Less}_{\mathcal{S}}$, where $\mathcal{S}$ is $\mathbb{N}$, $\mathbb{Z}$, $\mathbb{Q}$, or $\mathbb{R}$. This is done to make clear that, an order relation is a set of ordered pairs of elements of the given structure. It is important to keep this in mind, but with more structures to consider and a more detailed discussion to follow, keeping the formal notation absolutely correct all the time becomes a hindrance to reading (and writing). From now on, in accord with the usual mathematical practice, we will be taking liberties with notation. In particular, we will go back to the usual symbol $<$ for ordering in structures. When I say that $(A, <)$ and $(B, <)$ are ordered sets, I mean that in the first structure $<$ is a set of ordered pairs of elements of the set A, and the second, $<$ is the set of ordered pairs of elements of B. We will also use $x \leq y$ to abbreviate $x < y \vee x = y$. Moreover, in first-order statements we will no longer use atomic formulas $L(x, y)$, and just write $x < y$ instead. Sometimes it can lead to confusion, but it is a necessary compromise.

Dedekind cuts in an arbitrary linearly ordered set $(A, <)$ can be defined the same way as we did it for $(\mathbb{Q}, <)$, i.e. D is a Dedekind cut in $(A, <)$ if D is nonempty subsets of A, such that $D \neq A$, for all a in D, if $b < a$, then b is in D, and D has no largest element. A Dedekind cut D of $(A, <)$ is *rational* if there is an a in A, such that $D = \{x : x < a\}$. Otherwise D is *irrational.*

It is easy to see that $(\mathbb{N}, <)$ and $(\mathbb{Z}, <)$ have no Dedekind cuts.

Now comes a crucial definition.

Definition 5.1 A linearly ordered set $(A, <)$ is *Dedekind complete* if all Dedekind cuts of $(A, <)$ are rational.

As we have seen, $(\mathbb{Q}, <)$ is not Dedekind complete. It turns out that $(\mathbb{R}, <)$ is, and this has important consequences. One can think of the irrational Dedekind cuts as objects filling in all irrational gaps in linearly ordered sets—they *complete* the ordering of $\mathbb{Q}$.

From the Dedekind incomplete $(\mathbb{Q}, <)$ we constructed a larger ordered set $(\mathbb{R}, <)$. This set is larger, so potentially it could have new irrational Dedekind cuts. Then, one could move to a larger structure, in which these cuts become rational, but this new structure could also have new irrational cuts that would need to be filled in a larger structure, and so on. Such a process could have no end, but in the case of $(\mathbb{R}, <)$ it does end abruptly after just one step. It turns out that all Dedekind cuts in $(\mathbb{R}, <)$ are already rational (in $(\mathbb{R}, <)$!). This fact is not difficult to prove, but a proof requires more technicalities, and I do not want to make the text any heavier with notation than it is already. The more mathematically inclined reader may want to try to think how to prove it. The key is that rational numbers are dense in the set of real numbers, i.e. between any two real numbers, rational or not, there is always a rational number.

The ordering of $\mathbb{Q}$ is dense and has no largest and no smallest element. Moreover, $\mathbb{Q}$ can be constructed step-by-step. Cantor proved that any ordered set $(D, <)$ that is dense, has no largest and no smallest element, and can be constructed step-by-step, is an isomorphic copy of $(\mathbb{Q}, <)$. This means that there is a function f that to every rational number p assigns exactly one element $f(p)$ of D; every element of D is assigned to some p in $\mathbb{Q}$, and f preserves the order, i.e. for all p and q in $\mathbb{Q}$, $p < q$ if and only if $f(p) < f(q)$.

5.3.2 Summary

Let us summarize. We have just seen a construction of a new structure

$$(\mathbb{R}, \mathsf{Less}_{\mathbb{R}}, \mathsf{Add}_{\mathbb{R}}, \mathsf{Mult}_{\mathbb{R}})$$

with the domain $\mathbb{R}$ and the set of three relations on it. After identifying the rational numbers with the rational Dedekind cuts (in $\mathbb{Q}$), we can see that the new structure extends $(\mathbb{Q}, \mathsf{Less}_{\mathbb{Q}}, \mathsf{Add}_{\mathbb{Q}}, \mathsf{Mult}_{\mathbb{Q}})$, in the sense that the ordering, addition, and multiplication on $\mathbb{Q}$ do not change when we pass to $\mathbb{R}$. Moreover, the rational numbers $\mathbb{Q}$ form a dense subset of $\mathbb{R}$, and the ordered structure $(\mathbb{R}, \mathsf{Less}_{\mathbb{R}})$ is Dedekind complete. The last two properties are crucial.

Let us go back to geometry for a while. In geometry, basic objects, such as points and lines, are often treated axiomatically. The axioms of geometry do not explain what these objects are; they just tell us how they interact. For example, an axiom can say "any two distinct points lie on exactly one line."

Think of a geometric line as an infinite measuring tape. It has a point marked by 0, and points marking the integers at equal distances between the consecutive ones. It is an infinitely precise tape, meaning that it has points marking all rational numbers. Between any two such points, there is another point. So there is a dense set of points marking all rational numbers, and there still is room for points representing irrational numbers such as $\sqrt{2}$ and many other. This is all fine, but if we want to understand how the geometric line is actually *made* of points, simple geometric intuition does not suffice.

What we have done so far is this. Starting with the idea of counting we constructed the number systems of natural, integer and rational numbers. Each new structure, except the real numbers, was formally defined in terms of first-order logic in the previously constructed one. Moreover, we saw that each of the sets $\mathbb{N}$, $\mathbb{Z}$, and $\mathbb{Q}$ was built in step-by-step process. Even if one is not convinced that infinite processes can actually be completed via the magic limit step ω, these constructions can be carried out forever, producing larger and larger fragments of the intended structures. In this sense, we can say that we understand how those structures are built. We can almost see them. The situation changes dramatically when turn to $(\mathbb{R}, \mathsf{Less}_{\mathbb{R}}, \mathsf{Add}_{\mathbb{R}}, \mathsf{Mult}_{\mathbb{R}})$. This structure is also defined formally, but the definition is of a different character, and there is no step-by-step procedure in sight. There will be more about it in the next section, but let us concentrate on the positive aspects of the construction first.

We defined the real numbers as Dedekind cuts. These cuts—rather complex objects in themselves—are the elements of the new structure. A structure is a set with a set of relations on it. The set is $\mathbb{R}$. Each element of $\mathbb{R}$, as a Dedekind cut, and as such it has its own structure, but in the arithmetic structure of the real numbers it only matters how the elements of the domain are related to one another by the ordering, addition, and multiplication. Notice however, that we did use the structure of individual Dedekind cuts to define those relations.

The ordering of the reals is Dedekind complete; there are no gaps in it. One can say that $(\mathbb{R}, <)$ fulfills all expectations one has of the continuous geometric line. We defined a linearly ordered set that has the properties that we think a geometric line should have. We have built a model, an explicitly defined representation. Moreover, not only $(\mathbb{R}, <)$ provides a model of a geometric line, $\mathbb{R}$ is also equipped with the arithmetic structure given by $\mathsf{Add}_{\mathbb{R}}$, and $\mathsf{Mult}_{\mathbb{R}}$, so this justifies referring to the points on that line as numbers.

5.4 Dangerous Consequences

The construction of the real line $\mathbb{R}$ involves *all* Dedekind cuts, and because of that it is not really a construction in any practical sense. We can not perform infinitely many operations, but at least, if a the structure is build step-by-step in a well-defined manner, we can imagine the process continuing forever. No such conceptual comfort is available to describe filling in the gaps in the line made of rational points. It is a mathematical construction of a set-theoretic nature. For now, let us just accept that the real number line has been constructed, all gaps filled, and it can serve as an ideal measuring tape from which both the planar and the three dimensional coordinate systems can be constructed. We still do not quite understand how a line is made of points, but we at least have a chance to ground elusive geometric considerations in a seemingly more solid domain of sets. Unfortunately, it is not that simple. Some new and rather unexpected problems arise and they need to be analyzed carefully.

Consider two circles centered at the same point, one with radius 1 and the other with radius 2. The circumference of the first one is 2π, and the circumference of the second is 4π, so the second circle is twice as large as the first. For each point on the smaller circle, draw a line from the center passing through that point (see Fig. 5.2). The line will touch the larger circle at one point. You can also draw a line starting from a point on a larger circle and the line through this point and the center will touch the smaller circle at one point. We are using the fact that both circles are continuous curves hence there are no gaps in them. If we choose distinct points on the smaller circle, the corresponding points on the larger circle will also be distinct, and the same happens if we choose distinct points on the larger circle, the corresponding point on the smaller one are distinct. This reveals an interesting fact: there is a one-to-one correspondence between the set of points of the smaller circle and the set of points of the larger one. In other words, even though their sizes are different, both circles have the same *number* of points. We have not yet said what could be the "number" of points of an infinite set, but we have used the following plausible principle: if there is a one-to-one correspondence between elements of sets, then the number of elements in those sets should be the same. How is it that sets of the same size (the number of elements), can have different measures? We will not discuss it in detail, but let me just say that the main reason is that while the number of the elements of an infinite set, as defined by Cantor, is infinite, the measure of a set is a (finite) real number.[3]

In 1924, two Polish mathematicians, Stefan Banach and Alfred Tarski, published the following theorem. A solid ball in 3-dimensional space can be decomposed into five pieces. Each of the pieces can be rotated and shifted without changing distances between its points, and by such moves the pieces can be reassembled to form two

[3]Real numbers understood as Dedekind cuts, or, in the usual representation, as sequences of digits, are infinite objects. Here we mean that real numbers are finite in the sense that they measure finite quantities. For example, the area of a circle with radius 1 is π. It is definitely finite (less than 4), but π is a real number with an infinite decimal representation.

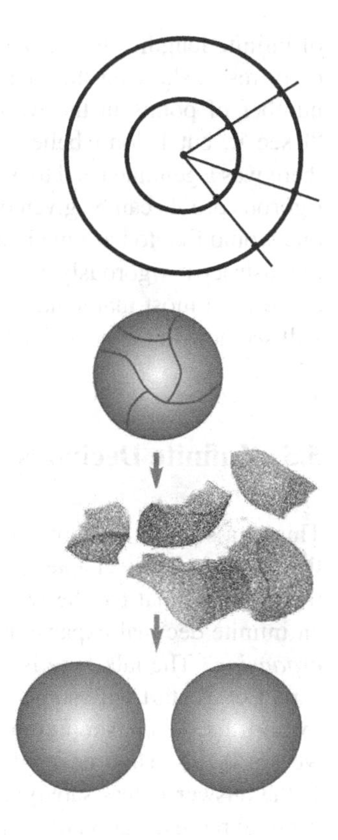

Fig. 5.2 One-to-one correspondence between points of a smaller and a larger circles

Fig. 5.3 Banach-Tarski paradox

solid balls, each of the same radius as the initial ball (see Fig. 5.3). In other words, one can cut a billiard ball into five pieces, and glue those pieces together to get two balls of the same size as the initial one. This result is known as the Banach-Tarski paradox. What is going on here is well-understood, and the result does not contradict anything we know about sets and geometry.[4] The explanation rests on a deeper understanding of properties of sets of (large) infinite size. If the pieces used in the Banach-Tarski decomposition are subjected to rigid motions in space, then their measures should not change. How can it be then, that the five pieces put together in one way make one ball, and in another two balls of the same size? The conclusion is that those pieces are not measurable. To be measurable is a technical term of measure theory. If arbitrary sets of points are admitted as objects, then, under some assumptions, there always will be objects that cannot be assigned a measure. Such are the pieces in the Banach-Tarski decomposition, but simpler examples of non-measurable sets had already been discovered earlier. One example, the Vitali set, is in the exercises to this chapter.

By the time Banach and Tarski proved their theorem, many other seemingly paradoxical phenomena had been discovered and studied. There is a bounded curve

[4]For full details see [37].

of infinite length. There is a continuous line completely filling a square. About an early result showing that the number of points in a unit circle is the same as the number of points in the whole 3-dimensional space, Cantor wrote to Dedekind: "I see it, but I don't believe it!" All those developments strongly indicated that there was a genuine need to base mathematics on solid axiomatic foundations. Since rigorous proofs can be given of results strikingly contradicting geometric intuitions, one would like to base mathematics on self-evident, undeniable truths, and then try reconstruct it rigorously just from those basic truths (the axioms). Not everyone agrees, but most mathematicians believe that task has been accomplished, and we will see how in the next chapter.

5.5 Infinite Decimals

This is a section for those who may wonder how decimal numbers are related to the subject of this chapter. In a college textbook we read: "The real numbers are the numbers that can be written in decimal notation, including those that require an infinite decimal expansion." What does that mean? What is an *infinite decimal expansion*? The talk here is about something actually infinite, so we prick up our ears. Let us start with some examples. The long division algorithm can be applied to find the decimal representation of any rational number. By dividing 8 into 1, we get 0.125. This means that $\frac{1}{8} = \frac{1}{10} + \frac{2}{100} + \frac{5}{1000}$. If we try to divide 3 into 1, the answer is less satisfying. Long division never ends, and we declare that the decimal representation of $\frac{1}{3}$ is the *infinite* decimal 0.3333333..., it is 0 followed by an infinite string 3's. But what could it possibly mean? The simple statement $\frac{1}{3} = 0.3333333\ldots$ stands for something rather advanced. It expresses that the infinite sequence of decimal fractions $\frac{3}{10}, \frac{33}{100}, \frac{333}{1000} \cdots$ *converges* to the number $\frac{1}{3}$. Each of these numbers is smaller than $\frac{1}{3}$, but for each number r less than $\frac{1}{3}$, no matter how close to $\frac{1}{3}$, there is a number in that sequence that is larger than r.

The infinite sequence 0.3333333.... is an example of a *repeating* decimal. It is often denoted by 0.(3), or $0.\bar{3}$, indicating that 3 repeats endlessly. If you randomly pick a fraction, there is a good chance that it is a repeating decimal. For example, $\frac{1}{7} = 0.142857142857 \cdots = 0.(142857)$. Even more interesting is $\frac{1}{17}$. Its decimal representation is 0.(0588235294117647). The number 0.1234343434... =0.12(34) is also an example of a repeating infinite decimal. It is eventually repeating.

Here is an interesting fact: every repeating infinite decimal converges—in the sense described above—to a rational number. For example, let us see what 0.(123) converges to. Let x be the real number to which the decimal 0.(123) converges.

$$x = 0.123123\ldots$$

Then, it follows that $1000x$ converges to 123.123123.... Since 123.123123... = 123+0.123123..., we get:

$$1000x = 123 + x$$

This equation can be easily solved, giving us the answer $x = \frac{123}{999}$. This fraction can be reduced. It is equal (equivalent) to $\frac{41}{333}$.

If you are familiar with the kind of algebra that we used above, a moment's reflection will convince you that the procedure just described works for any eventually repeating decimal. This is interesting in itself, but there is also a consequence that is actually quite fascinating. Since $\sqrt{2}$ is irrational, its decimal representation is not repeating. The infinite decimal representation of $\sqrt{2}$ always changes its pattern. It never runs into a loop of repeating digits. The fascinating part of this conclusion is that, even though we can only survey finite parts of the infinite sequence of digits in the representation of $\sqrt{2}$, we do know something about the whole complete infinite sequence. We know for sure something that we could never verify by direct checking.

How does the decimal expansion of $\sqrt{2}$ relate to the Dedekind cut representing this number? Recall that the cut representing $\sqrt{2}$ is $D_{\sqrt{2}} = \{p : p \in \mathbb{Q} \wedge p^2 < 2\}$. This representation gives us an effective procedure for finding the decimal expansion of $\sqrt{2}$. By direct calculations, one can check that

- 1 is the largest integer whose square is less than 2;
- 1.4 is the largest two digit decimal whose square is less than 2;
- 1.41 is the largest three digit decimal whose square is less than 2;
- 1.414 is the largest four digit decimal whose square is less than 2;
- 1.4 142 is the largest five digit decimal whose square is less than 2;
- and so on...

Proceeding this way, one can accurately compute arbitrarily long decimal approximations to $\sqrt{2}$. Of course, to say that one can compute, is a stretch. After a few steps, such calculations become too tedious to perform. The algorithm described above is not efficient, there are much better algorithms.

Exercises

Exercise 5.1 * *Prove that the ordered set* $(\mathbb{R}, \mathsf{Less}_{\mathbb{R}})$ *is Dedekind complete.*

Exercise 5.2 * *Write a formal definition of multiplication of real numbers represented as Dedekind cuts.*

Exercise 5.3 *We know that* $\sqrt{2}$ *is irrational. Prove that* $\sqrt{2}+1$ *and* $2\sqrt{2}$ *are also irrational. Hint: Assume that* $\sqrt{2}+1 = \frac{p}{q}$, *where p and q are integers, and use simple algebra to derive a contradiction; similarly for* $2\sqrt{2}$.

Exercise 5.4 *Prove that if a and b are rational numbers and* $a \neq 0$, *then* $a\sqrt{2}+b$ *is irrational.*

Exercise 5.5 *Give an example of two irrational numbers a and b such that* $a+b$ *is rational. Hint: For all a,* $a + (-a) = 0$.

Exercise 5.6 *Give an example of two irrational numbers a and b such that $a \cdot b$ is rational. Hint: For all $a \geq 0$, $\sqrt{a} \cdot \sqrt{a} = a$.*

Exercise 5.7 *Use long division to find the decimal representations of $\frac{1}{7}$ and $\frac{1}{17}$.*

Exercise 5.8 * ***The Vitali set.*** *In this exercise you are asked to fill in details of the construction of a set of real numbers that is not measurable.*

- *We define a relation E on the set of real numbers in the interval $[0, 1] = \{x : x \in \mathbb{R} \wedge 0 \leq x \leq 1\}$ by saying that the numbers a and b are related, if the distance between them is a rational number. Show that E is an equivalence relation.*
- *For each equivalence class $[a]_E = \{b : aEb\}$, we select one element c_a in it. The Vitali set V is the set $\{c_a : a \in [0, 1]\}$.*
- *For each rational number $p \in [0, 1]$ we define the set V_p to be the union of two sets $\{x + p : x \in V \wedge x + p \leq 1\}$ and $\{x + p - 1 : x \in V \wedge x + p > 1\}$. In other words, for each p, the set V_p is obtained by shifting the set V to the right by p, and cutting and pasting the part of it that sticks out beyond 1 at the left end of the interval $[0, 1]$.*
- *Here are two important properties of the sets V_p:*

  (1) *If $p \neq q$, then V_p and V_q are disjoint.*
  (2) *For each $x \in [0, 1]$ there is a rational number p such that either $x + p$ or $x - p$ is in V.*

- *Suppose the set V can be assigned some measure m. There are two possibilities: either $m = 0$ (yes, there can be nonempty sets with measure is 0, for example a single point, or a finite collection of points is like that) or $m > 0$. If $m = 0$, then, since V_p is obtained from V by rigid motions, the measure of V_p is also 0. It follows that the whole interval $[0, 1]$ is covered by the sets V_p, which are all disjoint, and whose measure is 0. The problem now is that the set of rational numbers can be constructed in a step-by-step process, and it follows that the whole interval $[0, 1]$ would be covered by disjoint sets of measure 0 in a step-by-step process. Measure theory does not allow this to happen, because then the measure of the whole interval $[0, 1]$ would be 0, and that is a contradiction. If $m > 0$, then by a similar argument, it follows that the interval $[0, 1]$ would be covered by infinitely many disjoint sets, each of measure m; hence the measure of $[0, 1]$ would have to be infinite, and that is a contradiction as well.*

Chapter 6
Set Theory

The ontological decision concerning infinity can then simply be phrased as: an infinite natural multiplicity exists.
Alain Badiou *Being and Event* [1]

Abstract In previous chapters we introduced mathematical structures, and we followed with a detailed description of basic number structures. Now it is time to look at structures in general. The classical number structures fit very well the definition: a set with a set of relations on it. But what about other structures? Are they all sets? Can a set of relations always be associated with them? Clearly not. Not everything in this world is a set. I am a structured living organism, but I am definitely not a set. Nevertheless, once a serious investigation of set theory got underway, it revealed a fantastically rich universe of sets, and it showed that, in a certain sense, every structure can be thought of a set with a set of relations on it. To explain how it is possible, we need to get a closer look at sets. As we saw in the previous chapter, the deceptively simple intuitive concept of set (collection) leads to unexpected consequences when we apply the well understood properties of finite sets to infinite collections. The role of axiomatic set theory is to provide basic and commonly accepted principles from which all other knowledge about infinity should follow in a formal fashion. There are many choices for such theories. In this chapter we will discuss the commonly used axioms of Zermelo and Fraenkel.

Keywords Axiomatic set theory · Axioms of ZF · Unordered pairs · Actual infinity · Power set axiom

6.1 What to Assume About Infinite Sets?

Sets are collections of objects. To construct a mathematical universe of sets, we will start with very little, strictly speaking with nothing—an empty set. Then we will determine how other sets can be constructed from it. What initially creates a difficulty in thinking about sets is that in the universe of sets, there is nothing but

R. Kossak, *Mathematical Logic*, Springer Graduate Texts in Philosophy 3,
https://doi.org/10.1007/978-3-319-97298-5_6

sets. There are only sets and nothing else. No numbers. No points. No triangles and circles. How can the rich world of mathematical objects and structures be recreated in such a setting? This is what we will see in this chapter.

Set theory as a mathematical discipline did not exist until the second half of the nineteenth century. Around 1870, Georg Cantor began a systematic study of certain sets of points on the number line, and, to do that, he developed a mechanism of counting beyond the finite. Cantor defined the notion of the size (cardinality) of a set and in the process of comparing sizes of various infinite sets he discovered a vast hierarchy of infinities. The new discipline attracted a lot of attention and one of the outcomes was the creation of the axiomatic system, known today as Zermelo-Fraenkel Set Theory, abbreviated by ZF. This formal theory does not explain what sets are. Instead, the axioms of ZF state that sets with certain properties exist, and that the universe of all sets is closed under certain operations, which means that those operations, when applied to sets, yield sets.

In formal theories, axioms can be chosen arbitrarily; they are often formulated just to see what their formal consequences are. For a theory as fundamental as set theory, the choice of axioms is a more demanding task. The axioms are all supposed to be evident. Some undoubtedly are, but certainly not all. A fuller coverage of all related issues would take us too far. Our present aim is just to get familiar with sets, set operations, and the set-theoretic language.

All axioms of ZF can be easily expressed without logical symbolism, but we will write them in a more formal way. The reason is to see how all statements about sets can be expressed in first-order logic. Some of the formal axioms are not easy to digest at first. We will be clarifying their meaning as we move along.

We will use the language of first-order logic with only one binary relation symbol $\in$. We will refer to this language as the *language of set theory*.

Since the axioms will express properties of sets, you should think of the variables $x, y, z, \dots$ as representing arbitrary sets. The choice of variables is not important. Later on, we will use other variables such as $A, B, \dots X, Y, \dots$. One can also use more descriptive names, like Set_1, Set_2, and the like. The membership relation $\in$ is binary, but for greater clarity, atomic formulas of the form $\in(x, y)$ will be written as $x \in y$. We read $x \in y$ as x is an element (or member) of y.

The symbol $\Longrightarrow$ is the implications symbol. In logic formalism, we write "if φ then ψ" as $\varphi \Longrightarrow \psi$. The symbol $\Longleftrightarrow$ is used for equivalence: $\varphi \Longleftrightarrow \psi$ abbreviates the statement $(\varphi \Longrightarrow \psi) \wedge (\psi \Longrightarrow \varphi)$.

Small finite sets can be given by a list of their members. The list of elements are traditionally written within curly brackets {, }. For example, the set whose only elements are a and b is $\{a, b\}$. We can give this set a name, say x. Then $x = \{a, b\}$. This notation is not part of our first-order formalism, but notice that the statement $(a \in x) \wedge (b \in x) \wedge \forall y[y \in x \Longrightarrow (x = a \vee x = b)]$ expresses that $x = \{a, b\}$, hence we can use the latter as an abbreviation.

Recall that if the same free[1] variable is used more than once in a formula, it stands in for the same set; however, when we use different variables, it does not automatically mean that the sets they refer to are different. For example, when we say that something holds "for all sets x and all sets y," we really mean "for all" and this includes the case when the x is the same as y.

We will start with the following three axioms. The first axiom declares that there is a set that is empty; it has no elements. The axiom of the empty set is not the most natural axiom that comes to mind when one thinks about sets. In fact, one could create a formal set theory without it, but, as we will see, it is convenient. The second axiom says that if x and y are sets, then there is a set z which contains all elements of the set x, all elements of the set y, and nothing more. We call such a set z the *union* of x and y and denote it by $x \cup y$. For example if $x = \{a, b\}$ and $y = \{a, c, d\}$, then $x \cup y = \{a, b, c, d\}$. The third axiom is the *axiom of extensionality*; we will comment on its importance below. Remember that all variables $x, y, z, \ldots$ represent sets.

Axiom 1 There is an empty set

$$\exists x \forall y \neg(y \in x)$$

Axiom 2 Unions of sets are sets

$$\forall x \forall y \exists z \forall t\ [t \in z \iff ((t \in x) \vee (t \in y))]$$

Axiom 3 Extensionality

$$\forall x \forall y\ [\forall z\ (z \in x \iff z \in y) \implies x = y]$$

Axiom 1 is a set existence axiom; it declares that a certain set exists. Axiom 2 is a closure axiom; it says that if certain sets exist, some other sets must exist as well. Axiom 3 is of a different nature. It declares that in the world of sets only the membership relation matters. If two sets x and y have exactly the same members, they must be equal. This axiom has important consequences. For example, it implies that there is only one empty set (a somewhat curious fact). Since there is only one empty set we can introduce a symbol for it; the one commonly used is $\varnothing$. Another consequence of Axiom 3 is that the sets $\{a, b\}$ and $\{b, a\}$ are equal. In specific examples of sets, linearity of notation forces us to list elements in a particular order. It follows from Axiom 3 that this order is irrelevant. Also, since the sets $\{a, a\}$ and $\{a\}$ have exactly the same elements, they must be equal. The expressions in curly brackets are not sets, they are lists of elements of sets for our informal use; if an element of a set is listed twice, it does not make the set any bigger.

[1] A free variable in a formula, is a variable that is not within a scope of a quantifier. For example, in $\forall x \exists y[x \in y \wedge \neg(z \in z)]$ only z is free.

There are more axioms. In fact there are infinitely many, and this is because some of them are in the form of axiom schemas. This means that a single statement is used to represent infinitely many statements, all of the same shape. This is the case of the next axiom, which is known as the *comprehension axiom*. The axiom says that if x is a set, and $\varphi(z)$ is a first-order formula with the free variable z, then there is a set y which contains exactly those elements of the set x which have the property expressed by $\varphi(z)$.[2]

Axiom 4 The axiom schema of comprehension

For every first-order formula $\varphi(z)$ in the language of set theory
$$\forall x \exists y\, [\forall z (z \in y \iff (z \in x \wedge \varphi(z))]$$

Axiom 4 is an *axiom schema* because we have one axiom for each formula $\varphi(z)$. It is known that we cannot reduce Axiom 4 to a single first-order statement in the language of set theory. Of course, as written it is a single sentence and it has the flavor of a first-order statement. It begins with "for all formulas $\varphi(z), \ldots$," so we are quantifying over formulas not sets. This obstacle can be overcome. Since (almost) all mathematical objects can be represented as sets, so can be formulas of set theory. There is no problem with treating the syntax of a first-order language in set-theoretic fashion. There is a much more serious problem though. To make sense of Axiom 4 in this way we would also need to be able to express set-theoretically the semantic part, i.e. we would need a set-theoretic definition of truth for formulas of set theory. In the second part of the book we will discuss the celebrated theorem of Alfred Tarski that implies that there are no such definitions.

To appreciate the strength of Axiom 4 we need to know what kind of properties can be expressed in the first-order language of set theory. Later we will see that almost everything about mathematics can expressed that way. For now let us consider one example. The property "the set z has at least two elements" can be expressed by the following formula with the free variable z: $\exists t \exists u[(t \in z) \wedge (u \in z) \wedge \neg(t = u)]$. Recall, that all variables represent sets; hence if z is an element of a set x, then z itself is a set, and it may or may not have at least two elements. Axiom 4 assures us that for every set x there exists a set y composed exactly of those sets z in x which have at least two elements.

The key step in Cantor's not-yet-formalized set theory, and in subsequent formalizations, was an open act of faith. This act is expressed by the *axiom of infinity*.

Axiom 5 The axiom of infinity

Infinite sets exist.

We are not saying here that "infinity exists," meaning that some processes can be potentially continued forever. The axiom says that there are sets that are actually infinite. There are many ways in which Axiom 5 can be expressed formally. We will

[2]Here, and in other axioms, we assume that formulas, such as $\varphi(z)$, can have other free variables.

examine how it is done in more detail, but before that we must introduce two more axioms.

Axiom 6 The axiom of pairing

$$\forall x \forall y \exists z \forall t\ [t \in z \iff ((t = x) \vee (t = y))]$$

Axiom 6 declares that any two sets x and y can be combined to form another set z, whose only elements are x and y. The set $z = \{x, y\}$ is called a *pair*, and sometimes we refer to is as an *unordered pair*. An important particular case is when $x = y$. Then the set $\{x, y\} = \{x, x\} = \{x\}$. Such a set is called a *singleton*.

The axiom of pairing looks innocuous. In the world of mathematics, pairing objects is one of the most elementary operations. In fact, we have already used it several times in our discussion of the number systems. From a philosophical viewpoint though, the self-evidence of the axiom becomes less clear. The universe of all sets is vast. One may think of it as a universe of all there is. What right do I have to take *any* object x and combine it with *any other* object y? What makes them a pair? Do x and y need to have something in common? According to the axiom, they don't. The axiom of pairing is a very peculiar axiom.

Axioms 6 looks very similar to Axiom 2. Compare the two axioms. The difference is not difficult to see. It can be proven that neither of the axioms implies the other.

Using the axioms listed so far, we can begin a reconstruction of familiar mathematical objects as sets. To make natural numbers, we will use a representation due to John von Neumann. Let us start with the number 0. In set theory, 0 can be represented by the empty set $\varnothing$. Notice that $\varnothing$ has 0 elements. The number 1 is defined as the set $\{\varnothing\}$. This set is not empty, it has exactly one element—the empty set. The next number 2 is $\{\varnothing, \{\varnothing\}\}$, and 3 is $\{\varnothing, \{\varnothing\}, \{\varnothing, \{\varnothing\}\}\}$. This rather mysterious process can be explained as follows. If we have a set x representing a natural number n, then the next number $n+1$ is represented $x \cup \{x\}$. This construction is designed so that the set representing a number n has exactly n elements. For each natural number n, we defined a unique set that represents it. We use the word "represents" metaphorically. We do not assume that the natural numbers *exist* somewhere else and here we reconstruct those numbers as sets. No, in set theory there is nothing but sets, and we *define* numbers as sets. Using the axioms we already listed, one can prove that all sets $\varnothing$, $\{\varnothing\}$, $\{\varnothing, \{\varnothing\}\}$, ... exist. Once this is done, we can give them the familiar names 0, 1, 2, ... To generate all those numbers (step-by-step) all we need is the empty set to begin with, and the rules that allow us to construct $x \cup \{x\}$ given x. This way we get set-theoretic representations of all natural numbers. We will need a bit more to define the set of all natural numbers.

A one way to express the axiom of infinity is to declare that there is a set containing all natural numbers as defined above. This is the meaning of the following formal statement

$$\exists x[\varnothing \in x \wedge \forall y(y \in x \Longrightarrow y \cup \{y\} \in x].$$

In this form, the axiom of infinity says that there is a set so large that it contains all natural numbers. To conclude that the set of natural numbers itself exists, more axioms of **ZF** are needed, but we will not go into the details. Let us just accept that the set of all natural numbers exists. As before, we will use $\mathbb{N}$ to denote the set of all natural numbers.

$\mathbb{N}$ is just a set, but it follows from axioms that the ordering relation on $\mathbb{N}$ is definable in set theory. In other words, if follows from the axioms that the set of ordered pairs $\{(m, n) : m \in \mathbb{N} \land n \in \mathbb{N} \land m < n\}$ exists. To this end, the relation $m < n$ must be defined in set-theoretic terms. This is done by observing that m is less than n if and only if there is a one-to-one function $f : m \longrightarrow n$ that is not onto. Here we take advantage of the fact that the set-theoretic natural numbers are sets, and that the axioms give us enough power to express statements about functions. All this development is a bit technical, but fairly routine. Once all of this is done we can define the ordered structure $(\mathbb{N}, <)$ and then declare that it is the first infinite counting number ω.[3]

The set-theoretic counting numbers have an interesting property: each number is the set of numbers smaller than itself. Thus $1 = \{0\}$, $2 = \{0, 1\}$, $3 = \{0, 1, 2\}$, and so on,[4] and ω can be identified with the ordered set of all finite counting numbers (i.e. all natural numbers). The great beauty of this approach is that there is no reason to stop at ω. The next counting number is $\omega + 1$, the one after that is $\omega + 2$ and so on. What is $\omega + 1$? According to the definition it is $\omega \cup \{\omega\} = \{0, 1, 2, \ldots, \omega\}$, and we extend the ordering relation by declaring that $m < \omega$ for all finite m.[5]

Starting with nothing (the empty set) and basic axioms that declare existence of other sets, one can generate a rich world of objects, which, while being "just sets" can be used to represent mathematical entities which do not, at first glance, look like sets at all. In fact, it turns out that, with very specific exceptions, most objects of modern mathematics can be interpreted as sets, and most theorems of mathematics can be derived as consequences of the axioms of set theory. This is absolutely remarkable. If you do not have a good idea of what "mathematical entities" other than numbers are, think of mathematics as the language of physics, which in turn provides models of our actual physical universe both in the micro and macro scales. All of those mathematical models can be interpreted as sets.

We have not listed all the axioms of **ZF** but the list is almost complete. A few axioms are missing, some are of a more technical nature and we will not discuss them here, but we do need one more—the Power Set Axiom—without which would

[3] Cantor defined ω to be the *order type* of $(\mathbb{N}, <)$.

[4] To see this, you need to look at the set theoretic representation of natural numbers. For numbers so constructed, it makes perfect sense to write expression like $3 \in 5$.

[5] A word of caution: What we described here is the process of generating transfinite counting numbers. They are known in set theory as *ordinal numbers* or just *ordinals*. Ordinal numbers are used to count steps in infinite processes, but they are not used to measure sizes of infinite sets. Cantor's *cardinal numbers* serve that second purpose.

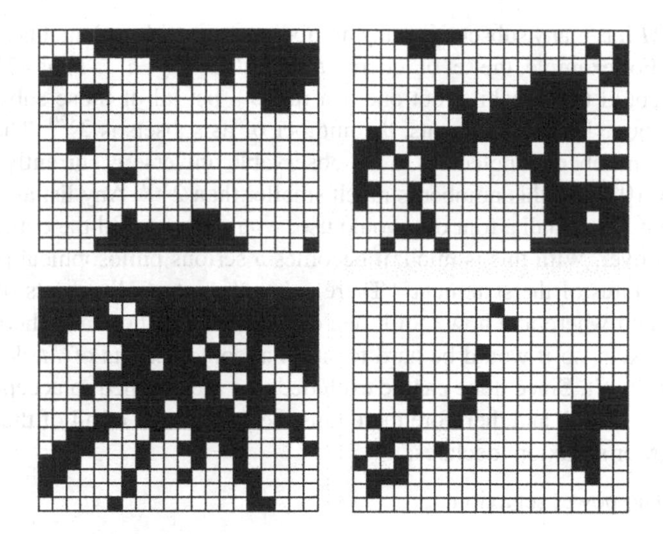

Fig. 6.1 Four of many subsets of a 320-element set

not be able to reconstruct the objects such as the number line, a plane, or the 3-dimensional space as sets.[6]

Let A be a set. Any collection of elements of A is called a *subset* of A. In particular, the empty set and the whole set A are considered subsets of A. Let A be a set of three different elements, let us call them a, b, and c. Here is a list of all subsets of A: $\varnothing$, $\{a\}$, $\{b\}$, $\{c\}$, $\{a, b\}$, $\{a, c\}$,$\{b, c\}$,$\{a, b, c\}$. The set A has three elements and eight subsets.

Imagine now the set M of pixels on a computer monitor contained in a rectangle of dimensions 20 pixels by 16 pixels. This rectangle is made of $20 \times 16 = 320$ pixels, so it is rather small (especially if your screen resolution is large). Think of pictures that are configurations of black and white pixels inside the square (see Fig. 6.1). Each picture can be identified with the set of black pixels (with the other pixels remaining white). Also, each subset of M can be identified with a picture by assigning black to all pixels in the subset and keeping all other pixels white. We see that the number of all possible black and white pictures inside the rectangle is the same as the number of subsets of the set M.

[6]Another word of caution: to say that we would not be able to reconstruct the real number line and other similar objects without the Power Set Axiom is not quite precise. We need that axiom to do it more or less naturally within the Zermelo-Fraenkel set theory. There are other axiomatic systems that do not have such an axiom, in which one can formalize much of modern mathematics, one of the more prominent ones being the second-order arithmetic. Those other systems are interesting, and they are studied for many reasons, but none of them has the status of ZF and its extensions that became a lingua franca of mathematics.

The set M has many subsets. For any number n, a set with n elements has exactly 2^n subsets. For example, the set of letters $\{a, b, c, d, e, f, g, h, i, j\}$ has $2^{10} = 1024$ subsets. It could take a while, but one can make a list all of those subsets. Since our set M above has 320 elements, the number of its subsets is 2^{320}. That number is huge. The number of molecules in the observable universe is currently estimated to be below 10^{81} and this number is much smaller than 2^{320}. Any list all subsets of M, no matter how small a font one would use, would have to fill the entire universe many times over. With this is mind, it becomes a serious philosophical problem to explain the status of the statement: "There exists the set of all subsets of M." If it does exist then where and how? Still, one can freely click here and there to select any set of pixels, so it would be hard to say that some subsets of M do exist, but some others don't. Since none can be excluded, we are inclined to accept that they all somehow "exist," and therefore there must "exist" the set of all of them. This is what the power set axiom declares:

Axiom 7 The power set axiom

For every set x, there exists the set of all subsets of x.

Following the route we outlined in Chap. 4, one can use the axioms to build the sets $\mathbb{N}$, $\mathbb{Z}$ and $\mathbb{Q}$ and the arithmetic operations as relations on them, but still need to see how relations are dealt with in set theory, and this is postponed until the next chapter.

Once we define $\mathbb{Q}$ as a set, the Power Set Axiom allows us to create the set of all its subsets. Then, using Axiom 4, one can show that the set of all Dedekind cuts exists. Then one can go on to define the set of all real numbers and their arithmetic operations, all in the language of set theory. This way, the real line, a classical object of mathematical analysis, gets reconstructed (from the empty set!) in set-theoretic terms. This is an important stepping stone in formalized mathematics, that opens the door to many further developments.

Exercises

Exercise 6.1 *Write down the set theoretic representations of* 4 *and* 5.

Exercise 6.2 *The first-order statement expressing that y is a subset of x is*

$$\forall z(z \in y \Longrightarrow z \in x).$$

Use this to write a first-order statement expressing the power set axiom.

Exercise 6.3 * *In set theory,* 1 *is defined as* $\{\varnothing\}$ *and* 2 *as* $\{\varnothing, \{\varnothing\}\}$. *Use the axioms of* ZF *to prove that* 1 *is not equal to* 2.

Exercise 6.4 *The set-theoretic representation of natural numbers in this chapter is due to John von Neumann. Earlier, Ernst Zermelo defined natural numbers this way:* $\varnothing, \{\varnothing\}, \{\{\varnothing\}\}, \{\{\{\varnothing\}\}\}, \ldots$. *Write the axiom of infinity using Zermelo's definition.*

Exercise 6.5 * *This exercise requires a lot of patience and some familiarity with formalized set theory. Write down formal set theoretic definitions of* $\mathbb{Z}$, $\mathbb{Q}$ *and* $\mathbb{R}$.

Part II
Relations, Structures, Geometry

Chapter 7
Relations

Abstract All further discussion will be based of a formal definition of relation, given in Definition 7.1. Then, in Definition 7.2, we introduce the central notion of definability in structures, and we proceed with examples of structures with very small domains, including the two element algebraic field F_2 presented in the last section.

Keywords Ordered pairs · Cartesian products · Relations · Definable sets · Fields · Structures with finite domains

7.1 Ordered Pairs

One more time: a structure is a set—the domain of the structure—with a set of relations on it.[1] To fully appreciate the concept, we delved into set theory. The universe of all sets is vast and mysterious. Since any set can serve as the domain of a structure, the universe of all structures is equally vast and mysterious. It contains the simplest structures, including the one whose domain is empty, and complex ones, such as the set of all Dedekind cuts with the relations of addition and multiplication. I need to stress a very important point. The structure on a given domain is determined solely by relations between its elements. Each set has its own "structure," namely the structure whose domain is the set, and whose only relations the membership relation $\in$. In most cases, this set-theoretic structure will now be disregarded. For examples, if the domain of a structure is the set $\{0, 1, \{1, 2\}\}$, then it is no longer relevant that $1 \in \{1, 2\}$, unless the membership relation $\in$ is one of the relations of the structure.

[1]In mathematics, a structure is usually defined as a set with a *sequence* of relations on it. That is an additional feature that allows to use the same relation as an interpretation for different relation symbols. We will not discuss such structures in this book, hence our definition is a bit simpler.

R. Kossak, *Mathematical Logic*, Springer Graduate Texts in Philosophy 3,
https://doi.org/10.1007/978-3-319-97298-5_7

Every set serves as the domain the trivial structure whose set of relations is empty.[2] Each trivial structure can be enriched by adding any set of relations.

In previous chapters, we appealed to intuitive understanding of relations based on familiar examples. Now we will deal with arbitrary relations, and we need to define this notion precisely.

We have already seen many relations, for example the order relation on the set of natural numbers $\mathsf{Less}_{\mathbb{N}}$, or addition and multiplication relations $\mathsf{Add}_{\mathbb{R}}$ and $\mathsf{Mult}_{\mathbb{R}}$ on the set of real numbers $\mathbb{R}$. Recall that $\mathsf{Less}_{\mathbb{N}}$ was the set of all ordered pairs on natural numbers (a, b) such that a is less than b. In the set-theoretic notation

$$\mathsf{Less}_{\mathbb{N}} = \{(a, b) : a \in \mathbb{N} \wedge b \in \mathbb{N} \wedge a < b\}.$$

The addition relation on $\mathbb{R}$ is the set

$$\mathsf{Add}_{\mathbb{R}} = \{(a, b, c) : a \in \mathbb{R} \wedge b \in \mathbb{R} \wedge c \in \mathbb{R} \wedge a + b = c\}.$$

The relations we have seen so far are certain sets of ordered pairs or ordered triples. In our discussion of the axioms of set theory we talked abut pairs $\{a, b\}$, and there was even an axiom declaring that all such pairs exist, but we also stressed that these are *unordered pairs*, i.e. $\{a, b\} = \{b, a\}$, for all a and b.

In relations, we want pairs to be ordered, but where does this order come from? Think of two arbitrary elements a and b, for example, a deck of cards and the moon. How are they ordered? One could alway say; I can consider the deck of cards first, and the moon later, so I am thinking of the ordered pair (a, b). Indeed, there is something very rudimentary in considering elements of a set in a particular order, so one would expect that perhaps in set theory order of pairs would be considered a fundamental notion, and that there would be some axioms expressing basic properties of ordered pairs such as: for all a and b there is an ordered pair (a, b); or for all a, b, c, and d, if $(a, b) = (c, d)$, then $a = c$ and $b = d$. This turns out to be unnecessary.

In set theory, ordered pairs can be defined in terms of unordered pairs. This can be done in several ways, and the one that is commonly used is due to the Polish mathematician Kazimierz Kuratowski. *Kuratowski's pair* of elements a and b, is the set $\{\{a\}, \{a, b\}\}$. The existence of such sets follows from the pairing axiom applied first to sets a and b to obtain the pairs $\{a, a\} = \{a\}$ and $\{a, b\}$, and then again to sets $\{a\}$ and $\{a, b\}$ to obtain $\{\{a\}, \{a, b\}\}$. As we suggest in the exercise section, the reader who wants to test his understanding of the axioms of ZF may want to prove the crucial property: if $\{\{a\}, \{a, b\}\} = \{\{c\}, \{c, d\}\}$, then $a = c$ and $b = d$. While the concept of ordered pair is important in foundations of mathematics, Kuratowski's definition does not reveal its "fundamental true nature." It is just an elegant technical

[2]Technically, the set of relations of a structure is never empty, since the equality relation symbol = is included in the first-order language and is interpreted in each structure as equality. Hence, to be precise, a trivial structure is a structure with one relation only—the equality relation.

device that allows us to limit the primitive notions of the axiomatic set theory to a bare minimum.[3] It is quite remarkable that just one relation $\in$ suffices for all set-theoretic needs.

Now we can reintroduce the notation: the ordered pair (a, b) is the set $\{\{a\}, \{a, b\}\}$. Since ordered pairs are sets, all other objects we will define using ordered pairs will be sets as well. We will also need ordered triples (a, b, c), quadruples (a, b, c, d), and in general arbitrary finite sequences of elements $(a_1, a_2, \ldots, a_n)$. All those objects can be defined in terms of pairs. For any three elements a, b, and c, the ordered triple (a, b, c) can be defined as the ordered pair $((a, b), c)$. Then, an ordered quadruple (a, b, c, d) can be defined as the pair $((a, b, c), d)$. We can continue, and in this way every finite sequence $(a_1, a_2, \ldots, a_n)$ can be represented as a set-theoretic pair, and consequently as a set.

7.2 Cartesian Products

For any sets A and B the *Cartesian product* $A \times B$ is the set of all ordered pairs (a, b) such that a is in A and b is in B. In the set-theoretic notation,

$$A \times B = \{(a, b) : a \in A \wedge b \in B\}.$$

For example, if $A = \{a, b\}$ and $B = \{1, 2, 3\}$, then

$$A \times B = \{(a, 1), (a, 2), (a, 3), (b, 1), (b, 2), (b, 3)\}.$$

The product $A \times B$ has six elements, which can be checked either by counting, or by arranging the pairs into a table with two rows and three columns (or three rows and two columns) which shows that the number of pairs is $2 \cdot 3 = 6$. This second method shows that, in general, if A has m elements and B has n elements, then $A \times B$ has $m \cdot n$ elements. This calculation will be relevant soon, when we explore the diversity of structures.

The *Cartesian square* of a set A, A^2, is the Cartesian product of a set with itself, i.e. $A^2 = A \times A$, so it is the set of all ordered pairs of elements of A. The Cartesian square is an example of a *Cartesian power*. The Cartesian cube, A^3, is the set of all ordered triples of elements if A, and in general, the n-th power A^n is the set of all sequences of length n of elements of A. In the set-theoretic notation,

$$A^n = \{(a_1, a_2, \ldots, a_n) : a_1 \in A \wedge a_2 \in A \wedge \cdots \wedge a_n \in A\}.$$

[3]The Wikipedia entry on ordered pairs has a list of other definitions and an informative discussion.

This definition also covers the case of $n = 1$. The first power of A^1 is the set of all sequences of length 1, which we can identify with single elements, hence A^1 is just A.

The axioms of ZF guarantee that all those Cartesian products and powers exist.

You may remember that much geometry and algebra is done using coordinate systems. The set of real numbers $\mathbb{R}$ is the real line, the set of pairs of real numbers is the coordinate plane $\mathbb{R}^2$, the three dimensional space is $\mathbb{R}^3$, and in general, going beyond school mathematics, $\mathbb{R}^n$ is the n-dimensional Euclidean space. Thanks to the set-theoretic construction of the real numbers, to Kuratowski, and to the axioms of ZF, all those objects can be considered as bona fide sets.

Cartesian products and sets of ordered pairs are basic concepts of modern mathematics, but their importance is often overlooked. The simple reason may be that they are in fact quite natural and intuitive. Why then, one could ask, do we need the sophisticated concepts such as Cartesian products, and devices such as Kuratowski's pair? Our goal is to clarify the concept of structure, but for now we seem to be obfuscating it with formal abstractions. Indeed, such a level of formality would be too high if we just wanted to create a language in which we could speak rigorously about familiar mathematical objects. However, here we want to do more. We want to understand structures in general. By making the definitions as general as we can, we aim at an all encompassing approach to make sure that no structure of interest is left behind. It is not a-priori clear that our set-theoretic approach will serve that purpose; however, one can make a strong argument that is does.

7.3 What Is a Relation on a Set?

We have already discussed many examples of mathematical relations, but strangely, we did not define the concept of relation in any formal way. To do this, we have had to wait until the set-theoretic machinery of Cartesian products has been introduced. This machinery will now allow us to give an elegant definition. The following mathematical examples serve as additional motivation. The examples are technical, and somewhat artificial, but I recommend that you spend some time thinking about them. The same examples will serve us to illustrate the idea of definability in structures.

Here are five relations on the set of real numbers $\mathbb{R}$.

- Relation R: a is twice as large as b.
- Relation S: b is between a and c.
- Relation T: On the number line, the distance between a and b is the same as between c and d.
- Relation U: Points (a, b), (c, d), and (e, f) in the coordinate plane $\mathbb{R}^2$ are collinear, i.e. they lie on the same straight line.
- The set V of all real numbers whose first nonzero decimal digit is 3. For example, the numbers, 3, 0.00003, and π are in V. Why V is a relation is explained below.

Each of our examples is a relation defined over the real numbers by some condition. We will take a closer look at those conditions in a moment, but for now the most important feature is that each of them can be represented as a subset of an appropriate Cartesian power of the set $\mathbb{R}$. When we think of relations, we often identify them with either properties or procedures that are used to verify whether some objects/elements are related or not. Now we will be defining them as sets of ordered pairs, ordered triples, and, in general, finite sequences of a fixed length. We have already applied this approach earlier when, for example, we defined the addition relation $\mathsf{Add}_{\mathbb{R}}$ on the set or real numbers to be the set of all triples (a, b, c) such that $a + b = c$.

- According to the definition of R, 2 is related to 1, 1 is related to 0.5, 0 is related to 0, and 2π is related to π. If we are given two numbers a and b, it is a simple matter to check if they are related or not; just multiply b by 2 and see if the result is a. Now instead of thinking of the description of R, we will consider R to be the set of all ordered pairs (a, b) such that $a = b + b$. In this sense R is a subset of $\mathbb{R}^2$ defined by the condition $a = b + b$.[4]
- Relation S can be identified with the set of all ordered triples (a, b, c), such that either $a < b$ and $b < c$, or $c < b$ and $b < a$.
- T relates pairs of ordered pairs of real numbers. The pair (a, b) is related to the pair (c, d) if the distance between a and b is the same as between c and d. Hence, T can be represented as a subset of $\mathbb{R}^2 \times \mathbb{R}^2$. However, we will do something a bit different. Instead of pairs of pairs $((a, b), (c, d))$, we can consider ordered quadruples (a, b, c, d) with no loss of information. Then, T can be represented as the set of all quadruples (a, b, c, d) such that the distance between a and b is the same as between c and d. In this representation, T is a subset of $\mathbb{R}^4$.
- Defining U involves an elementary algebra problem. Given three points (a, b), (c, d), and (e, f), one can write down an equation of the line through (a, b) and (c, d), and then check if the numbers e and f satisfy that equation. This gives us an equation in the variables a, b, c, d, e, and f, that defines U can as a subset of $\mathbb{R}^6$. The relation U is defined in terms of addition and multiplication of real numbers.
- The relation V deserves a separate comment. This example illustrates two points. We can identify V with the set of numbers that satisfy the condition defining V. Usually we think of relations between an object and another object, or relations between several objects; V is just a set of objects (numbers). It is not really a relation, in the common sense, but it is a relation according to Definition 7.1 below. It is a unary relation—a set of elements of the domain. The second point is about the definition of V. Given a decimal representation of a number, it seems straightforward to check if it is in the set V or not. Just look at the first nonzero

---

[4]There is a small issue here. If a is negative, then $2a$ is actually smaller than a, for example $2 \cdot (-1) = -2$, and -2 is smaller than -1 in the usual ordering of the real numbers. To make the definition of R more precise, we may say that b is "twice larger" than a if a and b are both positive, or both negative, and the absolute value of b is twice the absolute value of a.

decimal digits of the number, it is not that simple. The problem is that we talk here about all possible real numbers, and this includes all those incredibly small numbers whose decimal representations begin with 0.000..., and then continue with zeros for millions of decimal places, until the first nonzero digits appear. The set V has a simple description, but it follows from the results that we will discuss later, that, unlike all other relations listed above, V does not have a first-order definition in terms of addition and multiplication of real numbers.

The examples we just discussed were to prepare us for the general definition that is coming up now.

Definition 7.1 A *relation* on a set A is *any* subset of the Cartesian power A^n, for any $n > 0$.[5]

The examples R through V, are relations according to Definition 7.1. Hence $(\mathbb{R}, R, S, T, U, V)$ is a structure, and so is, for example, $(\mathbb{R}, R, T, V)$. Any selection of relations from the set $\{R, S, T, U, V\}$ determines a structure. There are 32 different ways to select a subset of a 5-elements set, including the empty set, so this gives us 32 different structures on the domain $\mathbb{R}$. Are all, or even some of those structures worth considering? It depends. Our examples are ad hoc, and they are not the most natural examples one would want to study, but still there are interesting mathematical questions one could ask about them. Are all those structures really different, and if so in what sense? Are some of the relations definable from other relations? Are all the relations definable in $(\mathbb{R}, \mathsf{Add}_{\mathbb{R}}, \mathsf{Mult}_{\mathbb{R}})$? We will answer some of these questions in the next section.

Before we finish this section, a few more words about terminology and a reminder about the important issue of separating syntax and semantics. None of the symbols used in this section belongs to the formal vocabulary of first-order logic. The names of sets and relations that we used are informal names of sets in the common language of mathematics. We use those names as abbreviations. The language of mathematics is rigorous and precise, but it is not formal.

To study structures using formal logic, the first step is to select relation symbols. It is a convenient, but often misleading practice to use the same letter for the informal name of a mathematical object, and the formal symbol that represents it. The tacit assumption is that the reader should be able to detect the difference from the context. Here is an example. Relation R, defined above, is a set of ordered pairs of numbers (a, b) such that $b+b = a$. To express that the numbers a and b are related, we can write $(a, b) \in R$. It is a statement of a mathematical fact written with symbols. It is informal. To study first-order properties of $(\mathbb{R}, R)$, we need a binary relation symbol, and we can use the letter R for it, but now understood as a formal symbol that can be used in formal formulas and sentences such as $\forall y \exists x R(x, y)$. This particular sentence is a true in $(\mathbb{R}, R)$, as it expresses the fact that a half of a

[5]One can define relation between elements of a set A and elements of a set B as an arbitrary subset of the Cartesian product $A \times B$, or in greater generality, a relation can be defined as an arbitrary subset of the Cartesian product of a finite number of sets $A_1 \times A_2 \times \cdots \times A_n$.

real number is a real number. The same formal sentence interpreted in $(\mathbb{Z}, Q)$, where Q is defined as the set of ordered pairs of integers (m, n) such that $m + m = n$, is false. A half of an integer m is an integer only if m is even.

7.4 Definability: New Relations from Old

Now, after crucial definitions have been introduced, we can begin to explore the world of structures. In various branches of mathematics, the study of either specific structures or the whole classes of them, has evolved into an art form involving advanced techniques applied at lofty levels of abstraction. What I am going to describe here is a small portion of those efforts that does not require anything too advanced. All we need is familiarity with the formalism of the first-order logic, in particular the important concept of definability that we have already discussed informally, and which now will be defined precisely. As we introduce more structures and other mathematical objects, we quickly run out of symbols to denote them. It is an old tradition in mathematical logic to use the German Fraktur font to denote structures. We will follow this tradition.

Definition 7.2 Let A be the domain of a structure $\mathfrak{A}$. A subset X of the Cartesian power A^n is *definable* in $\mathfrak{A}$ if there is a first-order formula $\varphi(x_1, \ldots, x_n)$ of the language of the structure such that

$$X = \{(a_1, \ldots, a_n) : \varphi(a_1, \ldots, a_n) \text{ is true in } \mathfrak{A}\}$$

First, we will take a closer look at definability in the structure of the real numbers with addition and multiplication as relations. Let $\mathfrak{R}$ denote the structure $(\mathbb{R}, \mathsf{Add}_{\mathbb{R}}, \mathsf{Mult}_{\mathbb{R}})$. $\mathfrak{R}$ is a classical mathematical structure, it is the *field of real numbers*.

A *field* is a structure $(F, \oplus_F, \odot_F)$, where $\oplus_F$ and $\odot_F$ are operations of "addition" and "multiplication" on the elements the set F satisfying all the basic properties of addition and multiplication of real numbers that you learned in school. In particular, F must have an element 0_F such that, for all a in F, $a \oplus_F 0_F = a$, and a different element 1_F such that for all a in F, $a \odot_F 1_F = a$. The real numbers with their addition and multiplication are a field, and so are the rational numbers, but the integers are not a field. The reason is that in a field every element, except 0_F, must have a multiplicative inverse, i.e. for every nonzero a in F, there must be a b such that $a \odot_F b = 1_F$. In $\mathbb{Z}$, $1 \cdot 1 = 1$ and $(-1) \cdot (-1) = 1$, so 1 and -1 have inverses, but all other integers don't.

An important note on notation: Our choice is to treat addition and multiplication as ternary relations. In the field of real numbers they are both subsets of the Cartesian cube $\mathbb{R}^3$. In the first-order language for $\mathfrak{R}$, we used two ternary relation symbols A and M to represent those relations. Then the atomic formula $A(x, y, z)$ is true in $\mathfrak{R}$, when the variables x, y, and z are interpreted by the numbers a, b, and c, respectively, if and only if $a + b = c$. This is an important point, that already has been stressed more than once. Keep it in mind, because now we are forced by the complexity of the formulas that will be involved, to go back to the more familiar notation. From now on, instead of $A(x, y, z)$, we will write $x + y = z$, and multiplication $x \cdot y = z$ instead of $M(x, y, z)$. Also, for the order relation, instead of $L(x, y)$, we will write $x < y$. Also, we will refer to $(\mathbb{R}, \mathsf{Add}_{\mathbb{R}}, \mathsf{Mult}_{\mathbb{R}})$ as $(\mathbb{R}, +, \cdot)$, and to $(\mathbb{R}, \mathsf{Less}_{\mathbb{R}})$ as $(\mathbb{R}, <)$.

In the set-theoretic definitions below, and later in the text, we will not explicitly refer to the domain of a structure (in this case $\mathbb{R}$), when it is clear form the context what the domain is. So, for example, instead of $\{a : a \in \mathbb{R} \wedge \varphi(a)\}$, where $\varphi(x)$ is some first-order property, we will just write $\{a : \varphi(a)\}$.

All relations R, S, T, and U that we defined in the previous section are definable in $\mathfrak{R}$, but V is not. Let us examine this more closely.

- Relation R is the set $\{(a, b) : a = b + b\}$. It is the set of pairs of real numbers that is defined by formula $x = y + y$. This shows that R is definable in $\mathfrak{R}$, but since the defining formula involves only addition R is definable already in the structure $(\mathbb{R}, +)$.
- Relation S is the union of two sets $\{(a, b, c) : (a < b) \wedge (b < c)\}$ and $\{(a, b, c) : (c < b) \wedge (b < a)\}$. This union is defined by formula $(x < y \wedge y < z) \vee (z < y \wedge y < x)$; hence it is definable in $(\mathbb{R}, <)$. It is also definable in $\mathfrak{R}$, but this is not immediately obvious. The order relation symbol $<$ is not included in the first-order language of $\mathfrak{R}$. The only relations of $\mathfrak{R}$ are the addition and multiplication. That the order relation is definable in $\mathfrak{R}$ follows from the fact that every positive number has a square root. For every positive real number a there is a real number b such that $a = b \cdot b$. All positive numbers are squares, and all non-zero squares are positive. It follows that the relation $a < b$ is defined by the property "there is a c such that $c \neq 0$ and $a + c^2 = b$." In other words, the formula

$$\exists z[\neg(z = 0) \wedge (x + z \cdot z = y)]$$

defines the order relation in the structure $\mathfrak{R}$. A small technical point: 0 is not a symbol of the language of $\mathfrak{R}$, so the formula above has one illegal symbol. It is not a serious problem though, since the set $\{0\}$ is defined in $\mathfrak{R}$ by the formula $x = x+x$; hence the formula defining the ordering can be written in the language with $+$ and $\cdot$ as the only non-logical symbols as follows

$$\exists z[\neg(z = z + z) \wedge (x + z \cdot z = y)].$$

- To define T, notice that the distance between points a and b on the number line is either $b - a$ if $a < b$ or $a - b$, if $b < a$. Then, the condition defining T can be written as $(a + d = c + b) \vee (b + d = a + c)$; showing that the relation is definable in $(\mathbb{R}, +)$.

- Relation U is more complex. One can use elementary algebra to show that its defining condition is equivalent to $d \cdot e + b \cdot c + f \cdot a = b \cdot e + a \cdot d + f \cdot c$, and this shows that U is definable in $\mathfrak{R}$.
- Finally, a few words about V. Its description is different from those of the previous examples. The definition is easy to understand, but in what formal language can it be expressed? Notice that it involves the higher level concept of decimal representation, and it does not seen to involve neither addition nor multiplication. In fact, V is not definable in $\mathfrak{R}$. The reason is that the set V is the union of infinitely many disjoint intervals. It includes the whole interval from 300 up to 400, not including 400, the interval from 0.03 to 0.04, not including 0.04, and infinitely many similarly defined intervals. It follows from a celebrated theorem of Alfred Tarski that no such set can be defined in $\mathfrak{R}$ by a first-order formula. Tarski's result was published by the RAND corporation in 1948 in a paper titled *A decision method for elementary algebra and geometry*[35]. The connection between first-order definability in structures and geometry is something that we will explore soon, but for a fuller discussion we need more examples and they will be provided in the next section.

7.5 How Many Different Structures Are There?

The smallest nonempty structures are the structures with one element domains. How many different structures with the domain $A = \{a\}$ can there be? A consequence of the formal Definition 7.1 is that in fact there are infinitely many. To define a structure with the domain $\{a\}$, we need to choose a set of relations. Since a relation on A is any subset of the Cartesian power A^n for any natural number n, while our choice is somewhat limited, we still have many options. For $n = 1$, A^1, which is A, has two subsets: the empty set, and the whole set A. For $n = 2$, A^2 also has two subsets:, the empty set of pairs, and the whole product $A^2 = \{(a, a)\}$, and similarly for all other n. For each n we have the empty set of sequences of length n, and the whole set $A^n = \{(a, a, \dots, a)\}$ consisting of one sequence of a's of length n. To define a structure, for each natural number n, we can either include the whole set A^n as a relation in A, or not. For example, (A, A^2, A^3, A^4) is a structure, and so is $(A, A^2, A^4, A^6, \dots)$ consisting of A and the set the Cartesian powers A^n for all even n. In general, for every set of natural numbers X there is a structure $\mathfrak{A}_X$, whose relations are the Cartesian powers A^n for all numbers n in X. If X and Y are different, then the corresponding structures $\mathfrak{A}_X$ and $\mathfrak{A}_Y$ are different. It follows that there are as many such structures as there are sets of natural numbers, and there is a vast multitude of such sets.

The structures $\mathfrak{A}_X$ are an example of mathematical trickery. It shows how to get many different structures out of almost nothing. Those structures are uninteresting, and it was nobody's intention to study such structures to begin with. Nevertheless, they deserve some attention. First of all, they illustrate an important aspect of formal methods. A formal concept is defined to capture an intuitive notion. It is designed to formalize crucial properties of familiar examples. When formalized, the concept often reveals additional features that may not have been intended. A formal definition says exactly what is says, and once it is in place it becomes the only criterion for qualifying whether an object falls under the defined category or not. Is empty set a structure? Sure, it is a set with an empty set of relations. Is the set with the set of all its subsets a structure? Sure, it is a set with all possible unary relations on it. The definition of structure is all-encompassing. It captures all kinds of structure. Whether it is good or bad, it depends. If a formal notion is too narrowly crafted to include all objects we intended to capture, it is clearly no good. If a notion is too general, it may allow exotic objects that can divert our attention from what we initially wanted to study, but this is a much lesser danger. In modern mathematics we aim at definitions that are as general as possible. We do not want to limit ourselves only to the objects we already know. The potential hidden in general definitions has often led to great mathematical discoveries.

The structures $\mathfrak{A}_X$ are not very interesting, they have no complexity in the sense that the only sets that are definable in them and the empty set, and the Cartesian powers A^n, for n in X. In other words, there is nothing more in those structures, than what we have put in there ourselves.

Let us now see that happens in larger domains. Let B be the two element set $\{a, b\}$. There are four unary relations on B: the empty relation $\varnothing$, one element relations $\{a\}$ and $\{b\}$, and the whole domain $\{a, b\}$. Any set of relations can be chosen to make a structure. There are $2^4 = 16$ different subsets of a four element set; hence we get 16 different structures made of unary relations on B.

Let us now count the number of different binary relations on B. A binary relation on B is any subset of the four element set of pairs $\{(a, a), (b, b), (a, b), (b, a)\}$. This set has $2^4 = 16$ subsets, hence there 16 different binary relations on B. Since any set of them can be chosen to make a structure, it follows that there are $2^{16} = 65{,}536$ different structures on B all made only with binary relations. The world of mathematical structures is vast!

There are lots and lots of structures even on very small domains, but are they all really different? For the domain B as above, consider $\mathfrak{A} = (B, \{a\})$ and $\mathfrak{B} = (B, \{b\})$. The structures are different as sets, but they do not look much different. Whatever can be said about $\mathfrak{A}$ can be also said about $\mathfrak{B}$. These two structures are isomorphic. They are really two different copies of the same structure. A clarification may be helpful here. One could object and say, wait, if one unary relation has a in it, and the other has b, the structures are different, and we just clearly stated the difference. The problem here is that when we say that $B = \{a, b\}$, we do not really mean it. Taken literally, $B = \{a, b\}$ could mean that B is the set of two letters, a, and b, but we do not mean that. The set B has two elements and we do not know what they are. They do not have names. I gave them names a and b, to facilitate the discourse. An alternative would be to say: Let $\mathfrak{A}$ be the set B together with a unary relation having one element in it, and let $\mathfrak{B}$ be the set B together with a unary relation with the other element of B in it. This is an important point. In the logic approach, what we initially know about the structure—what we assume is given to us—is only the domain and the relations on it. If you are given the trivial structure (B) with no relations, then you only know that the domain has two different elements and nothing more. Imagine that you see this structure and then, when you close your eyes, someone moves the two elements around. When you open your eyes, you will not be able to tell whether the elements have been moved or not. With no relations around, all elements of the structure are indiscernible.

The structures, $\mathfrak{A} = (B, \{a\})$ and $\mathfrak{B} = (B, \{b\})$, have domains with two elements, and each has one unary relation with one element in it. The permutation $f : B \longrightarrow B$ that moves a to b and b to a, is an isomorphism between $\mathfrak{A}$ and $\mathfrak{B}$. In model theory, it is a common practice to identify isómorphic structures. For a model theorist, $\mathfrak{A}$ and $\mathfrak{B}$ are not different at all. So the real question is not how many different structures are there on a given domain, but how many non-isomorphic structures there are. This is a much harder question that usually involves a much deeper mathematical analysis.

For an example of structures that are not isomorphic consider $\mathfrak{C} = (B, \{a\}, \{(a, b)\})$, and $\mathfrak{D} = (B, \{a\}, \{(b, a)\})$. Our relation symbols will be unary U and binary R. Let φ be the sentence

$$\exists x \exists y [U(x) \wedge R(x, y)].$$

Notice that φ is true about $\mathfrak{C}$, but false about $\mathfrak{D}$. In $\mathfrak{C}$ there is an element that has the property U, and that element is related to another element by the relation R; that element is a. In $\mathfrak{D}$ there are no such elements. It is important here that the binary relations on both structures are not symmetric,[6] and this lack of symmetry allows us to express a difference in a first-order way.

Here are visual representations of $\mathfrak{C}$ and $\mathfrak{D}$. Boldface font represents U and the arrow shows which element is related to which by R.

$\mathfrak{C}$ **a** $\longrightarrow$ b

$\mathfrak{D}$ **a** $\longleftarrow$ b

If colors were allowed in this book, we might have used a better visualization. In mathematical practice, we often think of unary relations as colorings. A unary relation is a set of elements of a structure, and it is convenient to think of the elements of that set as having been colored. Different colors, correspond to different unary relations. In the structures $\mathfrak{C}$ and $\mathfrak{D}$, a is colored, and b is not.

We showed that $\mathfrak{C}$ and $\mathfrak{D}$ are not isomorphic by finding a first-order order property (the sentence φ) which one structure has and the other has not. It is an interesting result in model theory that if two structures with finite domains are not isomorphic, then there is always a first-order property by which they differ.[7] This is no longer true for structures with infinite domains, and we will see examples later.

---

[6]A binary relation R is *symmetric* if for all a and b in the domain $(a, b) \in R$ if and only if $(b, a) \in R$.

[7]This theorem is proved in Appendix A.3.

7.5.1 A Very Small Complex Structure

Let us take a closer look at very special structure on a set of two elements. This time let the domain be $\{0, 1\}$, for reasons that will be clear in a moment. Since we have not yet defined any relations, 0 and 1 are just symbols.

A ternary relation on $\{0, 1\}$ is any subset of the Cartesian cube $\{0, 1\}^3$. The elements of $\{0, 1\}^3$ are sequences of 0's and 1's of length 3. Since there are 8 such sequences, there are $2^8 = 256$ ternary relations on $\{0, 1\}$. We will consider two of them:

$$A = \{(0, 0, 0), (1, 0, 1), (0, 1, 1), (1, 1, 0)\}$$

$$M = \{(0, 0, 0), (1, 0, 0), (0, 1, 0), (1, 1, 1)\}$$

We define our small complex structure to be $(\{0, 1\}, A, M)$, and we call it F_2. The names A and M are chosen for a reason. They are to be thought of as addition and multiplication. Look at A and think of the first two numbers in each triple as the numbers that are being added, and the third one as the result. For example, $(1, 0, 1)$ is in A; hence $1 + 0 = 0$. The addition that is defined this way does not differ from ordinary addition, with one exception: in F_2, $1 + 1 = 0$. Relation M represents multiplication, and it is just the ordinary multiplication restricted to 0 and 1.

Unlike other small structures we looked at in this chapter, F_2 is a genuine mathematical object. One can check that it is a field. All familiar properties of addition and multiplication that define fields—the field axioms—are valid in F_2. In particular, it implies that all algebraic rules that can be derived from the axioms, are also valid if F_2. For example, one of the consequences of the distributive property is the useful formula

$$(a + b) \cdot (a + b) = a \cdot a + a \cdot b + a \cdot b + b \cdot b.$$

Because it follows from the property that holds in all fields, it applies to all fields.[8] See Exercise 7.5.

F_2 is the smallest finite field, and it is not just a curiosity. It is an important algebraic structure. It has much to do with binary arithmetic, which in turn is the mathematical background for electronic computing.

[8] In elementary algebra this formula is usually written as $(a + b)^2 = a^2 + 2ab + b^2$.

Exercises

Exercise 7.1 *Use the axioms of* ZF *to prove that if* $\{\{a\}, \{a, b\}\} = \{\{c\}, \{c, d\}\}$, *then* $a = c$ *and* $b = d$.

Exercise 7.2 *Identify the axioms of* ZF *that are needed to prove that if* A *is a set, and* B *is a set, then* $A \times B$ *is also a set.*

Exercise 7.3 *Let* X *be set of natural numbers, and let* $\mathfrak{A}_X$ *be the structure defined in Sect. 7.5. Verify that the only sets that are definable in* $\mathfrak{A}_X$ *are the empty set, and the Cartesian powers* A^n, *for each* n *in* X. *Hint: According to Definition 7.2, if* $n \in X$, *then* A_n *can be defined as* $\{(a_1, \ldots, a_n) : a_1 = a_1\}$.[9]

Exercise 7.4 *Let* $A = \{a, b\}$ *be a two-element set. How many different structures with domain* A *can be made using only unary and binary relations? Hint: Find the numbers of subsets of* A *and* A^2, *and multiply them.*

Exercise 7.5 *Show that for all* a *and* b *in* F_2, $(a + b)^2 = a^2 + b^2$. *You can do it by direct checking, or better, by using the property* $(a + b)^2 = a^2 + 2ab + b^2$. *Hint: There is no* 2 *in* F_2, *but you can use* $1 + 1$ *instead.*

Exercise 7.6 *On the Cartesian plane* $\mathbb{R}^2$, *the circle with center at* $(0, 0)$ *and radius* 1 *is the set of points* (x, y) *such that* $x^2 + y^2 = 1$. *What is the set defined by this equation in* $F_2^2 = F_2 \times F_2$?

Exercise 7.7 *This is a "yes or no" question. Let* A *be a set. Are the structures* (A) *and* $(A, \varnothing)$ *isomorphic?*

[9]Instead of $a_1 = a_1$, we could use $a_i = a_i$, for any i, or in fact $a_i = a_j$ for any i and j.

Exercises

Exercise 7.1 [illegible]

Exercise 7.2 [illegible]

Exercise 7.3 [illegible]

Exercise 7.4 [illegible]

Exercise 7.5 [illegible]

Exercise 7.6 [illegible]

Exercise 7.7 [illegible]

Chapter 8
Definable Elements and Constants

Abstract In mathematics, or at least in the mathematics inspired by logical methods, to know a structure means to know all sets that are definable in it. In this chapter we will take a look at the smallest nonempty sets—those that have only one element. This a specialized topic, and it is technical, but it will give us an opportunity to see in detail what domains of mathematical structures are made of and in what sense they are "given to us."

Keywords Definable elements · Constants · The theory of a structure · Definable real numbers · Parametric definability

8.1 Definable Elements

In practice, mathematicians do not scrutinize elements of their structures in such a meticulous way as we will do in this chapter. Asking each time about the exact nature of elements would be counterproductive. Instead, new elements are introduced in definitions that are based on previously defined concepts. It can be a long process at the end of which it is easy to lose track of all the steps taken on the way. In our discussion of structures, we will adopt an even more radical approach. We will not be asking what mathematical objects are made of. We will accept that arbitrary sets as somehow given to us, and then, all other structure will gain concreteness, as they will be built, or at least represented, as sets.

We say that an element a of a structure is *definable*, if the set $\{a\}$, is definable. According to Definition 7.2, a set $\{a\}$ is definable, if there is a formula $\varphi(x)$ in the first-order language for the structure such that a is the only element of the domain that has the property expressed by $\varphi(x)$. Hence, the definable elements of a structure are those that can be identified by a first-order property. What can we say about individual elements of a structure?

R. Kossak, *Mathematical Logic*, Springer Graduate Texts in Philosophy 3,
https://doi.org/10.1007/978-3-319-97298-5_8

In our context everything is a set.[1] What are elements of sets then? They are also sets. So what are elements of those sets? Sets as well. Can it go on like this forever; are there sets in sets, in sets, in sets, and so on? No. One of the axioms of ZF—the *foundation axiom*—implies that there are no infinite sequences x_1, x_2, x_3, ..., such that x_2 is a member of x_1, x_3 is a member of x_2, and so on. The axiom of *foundation* implies that there are no such sequences, although there are axiom systems that allow them. Think of a set as an envelope that contains its elements. We open such an envelope, and look inside. Since everything is a set, unless the set is empty, inside there are sets. Think of those other sets as envelopes as well. If there are nonempty ones, let us take one and look inside. The foundation axiom assures us that this cannot go on forever. Proceeding in this way, we always have to reach a bottom after finitely many steps. So what are those elements (envelopes) at the bottom? They cannot have any elements, so they are empty. But there is only one empty set,[2] and this is what we find there at the bottom of each set. In set theory, the way it is designed, everything is built up from the empty set. No structure, other than the empty set, whose existence is declared by the axiom of the empty set, is assumed to exist a priori. Everything has to be created according to what the axioms are allowing us to create.[3]

The whole discussion above explains how structures are built in set theory. The process begins with the empty set, and laboriously other larger sets are constructed, and once a set is built, then by selecting a set of relations on it, a structure is made. Each element of a structure thus built has a structure of its own, but for our considerations now, the set-theoretic structure of each element must be forgotten. In analyzing a structure, we do not need to know what the elements of the domain are made of. All we are interested in are the properties of those elements with respect to one another, and the only properties that matter initially are those that can be directly expressed in terms of the relations of the structure. This is not as unnatural as it may seem at first glance. If you want learn about the structure of a city, first you identify the elements you want to pay attention to: streets, buildings, bridges, parks. We do not need to know what the buildings are made of, or whether the trees in a park have been recently trimmed. We just need a map showing how the elements spatially relate to one another. It would obfuscate the picture if one also wanted to include architectural plans for all buildings, or a complete book catalog of all libraries.

In our approach to structures, initially one treats all elements of a structure as equally significant, but one of first goals of the logical analysis is to recognize those elements that are distinguished by their first-order properties. In the next section we will examine some specific examples to see how it is done, but first we need another digression on how structures are given to us in the context of set-theory.

---

[1]In more advanced presentations of mathematical logic, even the syntactic objects are sets. Instead of thinking of formulas as marks on paper, we define them as certain sequences of sets.

[2]This is where the envelope metaphor brakes down. Clearly, there are many empty envelopes, but is follows from the axiom of extensionality that there is only one empty set.

[3]This is not quite accurate. The axiom of infinity also declares the existence of a set, but, as we have seen, the axiom postulates the existence of a set that contains infinitely many elements that are constructed from the empty set in a particular manner (the set-theoretic natural numbers).

8.2 Databases, Oracles, and the Theory of a Structure

How can we know anything about infinite structures at all? It is not a trivial matter. In this section, we will examine just one, not very precise metaphor hoping that it will shed light on the idiosyncrasies of mathematical thinking.

We defined the addition of real numbers as a relation. It is the set of all triples (a, b, c) such that $a + b = c$. We can think of this set as a database. If we want to know what is $a + b$, we do not need to perform any operations. In most cases it is not even possible. If a and b are real numbers with infinite non-repeating decimal representations, we do not even know how to begin. Instead of computing, one can think of searching the database until we find a triple (a, b, c) is in it. We can imagine doing this, but there is a problem. We tend to think of all real numbers as somehow given to us, but this is an illusion. They are infinite objects; there is very little we can actually do with them in an effective manner. So instead of "searching the database," let us think of an oracle, an infinite mind that can directly see all relations of a structure. Instead of searching, we will be asking the oracle. We could ask it to find the number c such that $a + b = c$. Since the oracle sees the whole database, it will instantly see the answer. But there is another problem: Can we even ask our question? Most real numbers are infinite objects, either as infinite strings of digits, or as Dedekind cuts. How do we inform the oracle what numbers we have in mind? We do have names for some special irrational numbers, such as $\sqrt{2}$, π, or e, but those names are meant for communication between us. Those numbers are defined in special ways, and for each such number there are computational algorithms giving us the digits of their decimal expansions. Look up π in Google (search "pi"). You will see pages and pages of decimal digits. Some webpages are live, they compute digit after digit in real time. To communicate with the oracle, we would need a mechanism for transmitting information with infinite content. There is no such mechanism. In general, we cannot ask direct questions about particular elements of the domain. In this sense it was not quite right, as we claimed earlier, to think of relations on the structure as given to us directly. They are not. They are set-theoretic objects. How can we know anything about them? This question has a complicated answer.

Most structures we are interested in are constructed for a particular purpose, and since we know how they are constructed, we know something about them. Often, but not always, what we know about a structure is first-order expressible. But here is another set-theoretic twist. Instead of considering particular first-order properties, we can make an audacious move, and also think of the totality of all first-order statements that are true about a structure. Let $\mathfrak{A}$ be a structure. The *theory* of $\mathfrak{A}$, denoted $\mathrm{Th}(\mathfrak{A})$, is the set of all first-order sentences true about $\mathfrak{A}$. It is the complete collection of all properties of $\mathfrak{A}$ that can be expressed in the first-order language with the symbols for all relations of $\mathfrak{A}$. Some structures, such as $(\mathbb{N}, +, \cdot)$, have immensely complicated theories, and we only know small fragments of them. For other, such as $(\mathbb{Q}, <)$, we have complete descriptions. In any case, whether we know the theory of a given structure or not, it is a well-defined set. It exists as a set-theoretic object, and we can think of it as a truth oracle for the structure.

Although first-order logic has not been introduced for this purpose, it turns out to be a perfect formalism for queries about databases. We will not go into details, but it is not difficult to convince oneself that any first order statement in the language of a structure can be interpreted as a database query. For example, if R is a binary relation symbol and R is a binary relation on a domain A, then, when given the sentence $\forall x \exists y R(x, y)$, the oracle looks at all a in A, and if for each such a she can find a b such that $(a, b) \in \mathsf{R}$ she can declare the sentence to be true, and false otherwise. Now, if we want to learn something about that structure we can ask any question we want, but if they are in the form of first-order sentences, then we understand well how a yes/no answer can be obtained by querying the database. We know the process.

If our goal is to understand a given structure, how do we know what first-order questions to ask? If we know nothing about the structure, then there is not much we can do. We need to know something in advance, based on how the structure was built and how its domain and its relations are defined. In such cases, it is often not that difficult to identify at least some definable sets. Usually one begins with identifying some special elements of the domain, and uses those to identify specific definable sets. For example, in all number structures that we discussed, 0 and 1 are of special importance, and they are definable (we will see how in the next section). From 1 one can define all integers, from those, solution sets to algebraic equations with integer coefficients.

To show that an element of the domain of a structure is definable, one has to find its definition, or to prove that such a definition exists. This is usually not a very hard task. It is much more difficult to prove that an element is not definable. We will need special tools to do that, and they will be described in Chap. 9. Now we will take a closer look at definability of real numbers.

8.3 Defining Real Numbers

In general, if $\varphi(x)$ is a formula in the first-order language for a given structure, the sentence $\exists x \varphi(x)$ is true in the structure, and $\exists x \exists y[\neg(x = y) \wedge \varphi(x) \wedge \varphi(y)]$ is false, it means that there is only one element in the domain that has the property expressed by $\varphi(x)$; this element is defined by the property $\varphi(x)$.

Recall that $\mathfrak{R}$ denotes the field of real numbers $(\mathbb{R}, +, \cdot)$. Consider the formula $x + x = x$. The only real number with that property is 0.[4] Hence, the formula $x + x = x$ defines 0 in $\mathfrak{R}$. The numeral 0 is not a symbol of the language of $\mathfrak{R}$, but since 0 is definable, we can introduce a formal symbol for it, and use it in first-order statements, because the added symbol can always be eliminated. For example, the

[4]Since we are dealing with an infinite domain, one could still ask how we know that. Our understanding of how the number system is constructed helps here. We do understand addition in $\mathfrak{R}$ well enough to know that much.

formula

$$\neg(x = 0) \wedge x \cdot x = x \qquad (*)$$

can be translated into

$$\neg(x + x = x) \wedge x \cdot x = x.$$

The only number that has the property expressed by formula $(*)$ is 1, so 1 is also definable. Once we have 1, we can define other natural numbers. For example, 2 is the only number x such that $1 + 1 = x$, hence 2 is definable. If a is a definable number, then $-a$ is definable as well as the only number x such that $x + a = 0$. This shows that all integers are definable. Rational numbers are also definable; $1/2$ is the only x such that $2 \cdot x = 1$, $2/3$ is the only x such that $3 \cdot x = 2$, and, in general, m/n is defined by $m = \cdot n$. Since all rational numbers are definable, in first-order formulas of the language of $\mathfrak{R}$ we can use names for all those numbers.

Every rational number is definable in $\mathfrak{R}$, but this does not imply that set of rational numbers is definable in $\mathfrak{R}$. Each rational number has its own individual definition, but those definitions cannot be combined to define $\mathbb{Q}$ as a set. We cannot combine those definitions because there are infinitely many of them. This does not rule out the possibility that there may be some other definition that works, but we do know that there can be no such definitions. This is a consequence of the already mentioned Alfred Tarski's analysis of definability in $\mathfrak{R}$.

An example of a number that is not definable is π, but it is not easy to prove.[5]

The ordering, i.e. the "less than" relation, is crucial for understanding the structure of the real numbers. We did not include a symbol for it in the language for $\mathfrak{R}$. That is because the ordering of real numbers is already definable in $\mathfrak{R}$. It is logically visible. The formula that defines the relation "x is less than y" is

$$\exists z[\neg(z = 0) \wedge x + z \cdot z = y].$$

Why does it work? If x and y are real numbers, and x is less than y, then the difference $y - x$ is a positive real number. Since every positive real number has a square root, the difference $y - x$ has a square root z, and for this z, $x + z \cdot z = y$. Since the ordering of the real numbers is definable, we can introduce the "less than" symbol $<$ and use it freely when expressing first-order facts about $\mathfrak{R}$.

In the argument above, we used subtraction $y - x$, but there is no symbol in the language of $\mathfrak{R}$ for it. There is also no symbol for division. The reason is that both these operations can be defined in terms of $+$ and $\cdot$. Indeed, for all numbers x, y, and z

$$y - x = z \Longleftrightarrow z + x = y$$

[5] It follows from a theorem proved in 1882 by Ferdinand von Lindemann (1852–1939). The theorem states that π is not a solution of any polynomial equation with integer coefficients (it is *transcendental*).

and

$$y/x = z \iff x \cdot z = y.$$

This all works well, with the exception of $x = 0$ and $y = 0$ in the second formula. If $x = 0$ and $y = 0$ were allowed, then $0/0 = z$ would be true for any number z. Notice that if x is zero and y is not, then the definition is still fine, because then $0 \cdot z = y$ for no number z, which means that division by zero is undefined.

With the aid of the ordering, we can now define more real numbers. For example, there are exactly two numbers x such that $x \cdot x = 2$. They are $\sqrt{2}$ and $-\sqrt{2}$. The formula $x \cdot x = 2$ defines the set $\{\sqrt{2}, -\sqrt{2}\}$, so it does not identify any of these two numbers uniquely, but the formula $x > 0 \wedge x \cdot x = 2$ defines $\sqrt{2}$, and $0 > x \wedge x \cdot x = 2$ defines $-\sqrt{2}$.

In a similar fashion, using algebraic equations, one can define many other real numbers, but one cannot define them all. The set of all first-order formulas is infinite, but small. It can be built in a step-by-step process. It is *countable*. The set of all real numbers is *uncountable*, it cannot be built step-by-step. It follows that not all numbers can have first-order definitions.

Talking about $\mathfrak{R}$ in a first-order way, we are allowed to use quantifiers that refer to the totality of all numbers, but, unless they are definable, we cannot refer to particular numbers. That does not mean that we cannot talk about undefinable numbers in a first-order way at all. In fact, we can, and we will, but to do it, we need to expand the language. For example, to talk about π, we can add to the language a new relation symbol $P(x)$, and interpret it as the set $\{\pi\}$. This gives us a new structure $\mathfrak{R}_\pi = (\mathbb{R}, +, \times, \{\pi\})$. π is not definable in $\mathfrak{R}$, but it is in $\mathfrak{R}_\pi$. A formula that defines it is $P(x)$. We can do more. For each real number r, we can add a new relation symbol $P_r(x)$ and interpret it in the set of real numbers as the set $\{r\}$. So now we have a new structure in which all real numbers are definable. It is a genuine first-order structure consisting of the set of real numbers, relations of addition and multiplication, and a huge infinite set of relations $\{r\}$, one for each real number r.

8.4 Definability With and Without Parameters

In everyday mathematical practice, one freely refers to arbitrary elements in structures. To do it in the first-order formalism, we expand the language of the structure so that we can use names for elements. In the previous section, we saw how it is done in the case of the field of real numbers. Let us see how it is done in general.

If an element a of the domain of a structure $\mathfrak{A}$ has a property that is expressible by a first-order formula $\psi(x)$, then we are tempted to say that $\psi(a)$ holds in $\mathfrak{A}$. The problem is that a is not a symbol in the language, and therefore $\psi(a)$ is not an expression of first-order logic. It mixes the two worlds: the syntactic and the semantic, and this is a serious violation. One could say; why don't we just add a symbol to the language to represent the element, and allow formal expressions involving the symbol? There is no problem if a is definable in $\mathfrak{A}$. Suppose that a is the only element of $\mathfrak{A}$ which has a first-order property $\varphi(x)$. If $\psi(x)$ is a formula of the language of the structure, then since a is the only element of the structure having the property $\varphi(x)$, the statement $\forall x[\varphi(x) \implies \psi(x)]$ expresses that a has the property $\psi(x)$. We can now introduce a new symbol s_a, and treat expressions such as $\psi(s_a)$ as abbreviations of $\forall x[\varphi(x) \implies \psi(x)]$. This way we can formally talk about definable elements without changing the formalism. We can do something similar for all elements of the domain, even if they are not definable.

For each element a of the domain of a structure $\mathfrak{A}$, we can add a new relation symbol $U_a(x)$ and interpret it in $\mathfrak{A}$ as the set $\{a\}$. Then $\mathfrak{A}$ expanded by adding all those new relations is a new structure in which each element is now definable. Indeed, each element a is defined by the formula $U_a(x)$. Then, to simplify notation, for each element a, we can introduce a constant symbol s_a, and treat expressions $\psi(s_a)$ as abbreviations of $\forall x[U_a(x) \implies \psi(x)]$. In this way, we can turn every structure $\mathfrak{A}$ into an expanded structure, which we will denote by $\overline{\mathfrak{A}}$, considered as a structure for the expanded language in which each element has a definition.

What is this formal fuss all about? In standard presentations of first-order logic, in addition to relation symbols, the syntax usually includes function symbols and constants. None of it is absolutely necessary, since functions can be considered as relations of a special type, and constants, as we have seen, can be treated as unary relations. For the presentation here, I have decided to use Occam's razor to reduce the formal apparatus to the minimum. In mathematical practice such frugality is not needed. In fact, one often considers structures in the first-order language including constants for all elements. The elements are treated as *parameters* and used freely in mathematical formulas. For structures whose domains are well-understood, such as the natural numbers or the rational numbers, it is not a problem at all, since for all elements we can use their standard representations as formal names. The approach is more problematic when applied to larger and more complex domains, such as the real numbers $\mathbb{R}$. Since the real numbers are uncountable, the set of their names has to be uncountable as well. This means that we can no longer claim that the set of all formulas of the extended language can be built in a step-by-step process. The set of formal expressions itself becomes a set-theoretic object, and it has to be investigated by set-theoretic means. There is no problem with doing all of that mathematically—modern mathematics is firmly based in set-theory—but notice that the distinction between syntax and semantics gets blurry, and we no longer stand on the firm (or firmer) ground, where the syntax is much simpler and better understood, than the structures it serves to describe. Another technical problem that has to be resolved is that if for a large domain we need a large set of constants, then where do those constants come from? Think of possible names for real numbers. Real numbers are represented by arbitrary Dedekind cuts, or finite or infinite sequences of digits. There is no rule or pattern for creating symbolic names for such objects, but there is an elegant way out. We can just declare that each element of a structure stands for its own name. After all, we never said what the symbols of first-order logic are supposed to be. Since there is no formal restriction, they can in fact be any kind of objects we like, so we can use elements of structures themselves. Why not? Technically, there is no problem, but the syntax vs. semantics division line gets even more difficult to demarcate.

In the next chapter, we will discuss some notions that are used to classify structures. They involve definability in languages that include names for all elements. Therefore let us finish this chapter with a formal definition. Compare it with Definition 7.2.

Definition 8.1 Let $\mathfrak{A}$ be a structure with domain A. For each natural number $n > 0$, a subset X of the Cartesian power A^n is called *parametrically definable* in $\mathfrak{A}$, if it is definable in the expanded structure $\overline{\mathfrak{A}}$.

Exercises

Exercise 8.1 *Write first-order formulas defining* 3, $\frac{1}{3}$, *and* $\frac{2}{3}$ *in* $\mathfrak{R}$, *using only symbols* $+$ *and* $\cdot$. *Hint:* $\frac{2}{3}$ *is a solution of the equation* $3x = 2$.

Exercise 8.2 *Write a first-order formula defining* $-\sqrt{2}$ *in* $\mathfrak{R}$, *using only symbols* $+$, $\cdot$, $<$, *and* 2. *Hint:* $\sqrt{2}$ *and* $-\sqrt{2}$ *are the only solutions of the equation* $x^2 = 2$.

Exercise 8.3 *Write a first-order formula defining* $\sqrt{2}$ *in* $\mathfrak{R}$, *using only symbols* $+$ *and* $\cdot$ *(no* $<$*). Hint: Recall that the ordering of the real numbers can be defined in* $(\mathbb{R}, +, \cdot)$.

Exercise 8.4 *Show that every element of the ordered set* $(\mathbb{N}, <)$ *is definable. Hint:* 0 *is the least element of* $\mathbb{N}$, 1 *is the least element of* $\mathbb{N}$ *that is greater than* 0, 2 *is the lest element of* $\mathbb{N}$ *that is greater than* 1, *and so on.*

Exercise 8.5 *Let* $\mathfrak{A}$ *and be a structure with domain A. Show that every finite subset of A is definable in the expanded structure* $\overline{\mathfrak{A}}$ *(see Definition* 8.1*).*

Chapter 9
Minimal and Order-Minimal Structures

Abstract Before we get into details regarding number structures, we will examine definability in cases that are easier to analyze. We define two important classes of structures: *minimal*, in Definition 9.1, and *order-minimal*, in Definition 9.4. The important concepts of *type* and *symmetry* were already introduced in Chap. 2; here we define them in general model-theoretic terms and use them to analyze the orderings of the sets of natural numbers, integers, and rationals.

Keywords Minimal structures · Order minimal structures · Types of elements in ordered sets · Symmetries of ordered sets · Order types of sets of numbers

9.1 Types, Symmetries, and Minimal Structures

For any structure $\mathfrak{A}$, any element of the domain is parametrically definable; hence all finite subsets of the domain are definable as well. For example, if a, b, and c are elements of the domain, then the formula $(x = a) \vee (x = b) \vee (x = c)$ defines $\{a, b, c\}$. If a set X is defined by a formula $\varphi(x)$, then its complement, i.e. the set of all elements of the domain that are not in X, is defined by $\neg\varphi(x)$. A set whose complement is finite is called *cofinite*. It follows that all finite and all cofinite subsets of the domain of any structure are parametrically definable. This brings us to the following definition.

Definition 9.1 A structure $\mathfrak{A}$ is *minimal* if every subset of its domain that is parametrically definable is either finite or cofinite.

Since all structures with finite domains are minimal, we will be only interested in structures with infinite domains. In a minimal structure, all parametrically definable subsets can be defined by formulas not involving the relation symbols of the language, other than $=$. Notice that the definition involves only subsets of the domain of the structure, and not subsets of its Cartesian powers. The definition cannot be made stronger by demanding that parametrically definable subsets of

R. Kossak, *Mathematical Logic*, Springer Graduate Texts in Philosophy 3,
https://doi.org/10.1007/978-3-319-97298-5_9

all Cartesian powers are also either finite or cofinite, because for structures with infinite domains it is never the case. For example, if A is infinite, then the subset of A^2 defined by the formula $x = y$ is neither finite nor cofinite.

In Chap. 2 we defined a notion of symmetry of a graph. Now we will generalize this definition to arbitrary structures. Informally, a symmetry is a rearrangement of the elements of a structure that does not change the way in which the elements are related. Recall that a permutation of a set A is a function $f : A \longrightarrow A$ that is one-to-one and onto.

Definition 9.2 Let $\mathfrak{A}$ be a structure with domain A. A permutation f of A is a *symmetry* of $\mathfrak{A}$ if for every relation symbol R of arity n in the language of $\mathfrak{A}$, for all $a_1, a_2, \ldots, a_n$ in A the sentence

$$R(a_1, a_2, \ldots, a_n) \Longleftrightarrow R(f(a_1), f(a_2), \ldots, f(a_n)) \qquad (*)$$

is true in $\overline{\mathfrak{A}}$.[1]

Symmetries not only preserve relations, they also preserve all definable properties. This is the content of the next theorem. It is a very useful tool for detecting undefinability of elements and sets.

Theorem 9.1 *Let $\mathfrak{A}$ be a structure with domain A and let f be a symmetry of $\mathfrak{A}$. Then, for any first-order formula $\varphi(x_1, x_2, \ldots, x_n)$ and all $a_1, a_2, \ldots, a_n$ in A the sentence*

$$\varphi(a_1, a_2, \ldots, a_n) \Longleftrightarrow \varphi(f(a_1), f(a_2), \ldots, f(a_n)) \qquad (**)$$

is true in $\overline{\mathfrak{A}}$.

We will use Theorem 9.1 to show that if X is a subset of A that is defined by a formula without parameters $\varphi(x)$, and f is a symmetry of $\mathfrak{A}$, then the image of X under f, i.e the set $\{f(a) : a \in X\}$ is X. Let us see why. This will be a routine argument often called "chasing the definitions."

If a is an element of X, then $\varphi(a)$ holds in $\overline{\mathfrak{A}}$, hence, because f is a symmetry, $\varphi(f(a))$ must hold as well, which means that $f(a)$ is in X. This shows that the image of X under the symmetry is contained in X. To see that it is all of X, suppose that b is in X. Because f is onto, there is an a in A such that $f(a) = b$. Because b is in X, $\varphi(b)$ holds, which means that $\varphi(f(a))$ holds, and this implies that $\varphi(a)$ holds. Now, since $\varphi(a)$ holds, a is in X and this concludes the argument.

A similar argument shows that if X is a subset of A defined by a formula with parameters $\varphi(x, b_1, b_2, \ldots, b_n)$, and f is a symmetry that fixes all parameters, i.e. $f(b_1) = b_1$, $f(b_2) = b_2, \ldots, f(b_n) = b_n$, then the image of X must be X. Hence, if for a given X and any finite set of parameters we can find a symmetry

[1] Symmetries of structures are known in mathematics as *automorphisms*.

that fixes the parameters and moves some element of X outside X, that will show that X is not parametrically definable in the structure. We will use this to establish non-minimality of certain basic structures.

In Chap. 2 we defined types of vertices in graphs, now we consider types in arbitrary structures.

Definition 9.3 If $a_1, a_2, \ldots, a_n$ is a finite sequence of elements of the domain of a structure $\mathfrak{A}$, then the *type* of $a_1, a_2, \ldots, a_n$ in $\mathfrak{A}$ is the set of all first-order formulas $\varphi(x_1, x_2, \ldots, x_n)$ such that $\varphi(a_1, a_2, \ldots, a_n)$ holds in $\overline{\mathfrak{A}}$.

The type of a finite sequence is the set of all its first-order properties. Types are easy to define, but it does not mean that they are easy to describe. The classification of all possible types realized in a structure is an important task. Sometimes all possible types can be easily classified according to some well-defined criteria, and we will see how it is done in a few cases. Sometimes such an analysis involves much hard work. Sometimes it cannot be achieved at all, and this is not because we do not know how to do it, but because we can prove that a classification of the kind we consider intelligible does not exist.

Theorem 9.1 can also be formulated as a statement about types: if $f : A \longrightarrow A$ is a symmetry of $\mathfrak{A}$, then for all $a_1, a_2, \ldots, a_n$ in A the type of $a_1, a_2, \ldots, a_n$ in $\mathfrak{A}$ is the same as the type of $f(a_1), f(a_2), \ldots, f(a_n)$. Symmetries preserve types.

9.2 Trivial Structures

Recall that a trivial structure consists of a domain with no relations. As one should expect, trivial structures are minimal. We will prove it. The proof is an argument about trivial structures, and it is rather trivial itself, but to a novice the details may not be that obvious. The argument uses Theorem 9.1 and some formal trickery. It is worth reading, as it shows the power of the methods that are the subject of this book. Paradoxically, when we get to discuss less trivial structures in the following section, the arguments may be easier to follow, so you may want to read about those other examples first.

Let A be the domain of a trivial structure $\mathfrak{A}$, and let $a_1, a_2, \ldots, a_n$ be a sequence of parameters from A. Let $\varphi(x, a_1, a_2, \ldots, a_n)$ be formula with one free variable x and the parameters as displayed, and let X be the subset of A defined by it. If X is a subset of $\{a_1, a_2, \ldots, a_n\}$, it is finite. Otherwise, it contains an element b outside the set of parameters. We will show that then X must contain all such elements; hence it is cofinite. So let us assume that b is not one of the parameters.

Since b is in the set defined by $\varphi(x, a_1, a_2, \ldots, a_n)$, $\varphi(b, a_1, a_2, \ldots, a_n)$ holds in $\overline{\mathfrak{A}}$. Let c be another element of A that is not in $\{a_1, a_2, \ldots, a_n\}$. Let $f : A \longrightarrow A$ be such that $f(b) = c$, $f(c) = b$, and for all other a in A, $f(a) = a$. Clearly, f is a permutation of A, and since $\mathfrak{A}$ has no relations, f vacuously satisfies the condition in Definition 9.2. Moreover, f is not only a symmetry of $\mathfrak{A}$. Since it fixes all parameters $a_1, a_2, \ldots, a_n$, it is also a symmetry of $(A, a_1, a_2, \ldots, a_n)$. Then, it follows from Theorem 9.1 that $\varphi(b, a_1, a_2, \ldots, a_n)$ holds in $\mathfrak{A}$, proving that c is in the set defined by $\varphi(x, a_1, a_2, \ldots, a_n)$, and proving our claim.

9.3 The Ordering of the Natural Numbers

Next, we will consider $(\mathbb{N}, <)$, consisting of the natural numbers with their ordering. $(\mathbb{N}, <)$ is minimal, but we will not prove it yet. This will be done in Chap. 12, where a powerful technique is described that will allow us to give a slick proof. For now, let us examine the types of elements in $(\mathbb{N}, <)$.

We have already seen that every natural number is definable in $(\mathbb{N}, <)$. Let us repeat the argument in a more general setting. If a and b are elements of an ordered set $(A, <)$, $a < b$, and there are no elements between a and b, then we call b the *successor* of a, and we call a the *predecessor* of b. If b is a successor of a, then b is the only element x of A such that the following holds in $(A, <)$:

$$a < x \wedge \forall y[a < y \Longrightarrow [(x < y) \vee (x = y)].$$

This shows that if b is a successor of a, then b is definable in $(A, <, a)$. It follows that if a is definable in $(A, <)$, then so is b. A similar argument applies to predecessors, hence, if b is a successor of a, then a is definable in $(A, <)$ if and only if b is definable.[2]

Since 0 is the least element of $\mathbb{N}$ and this property is expressible by a first-order formula, 0 is definable in $(\mathbb{N}, <)$. Since 1 is a successor of 0, 1 is definable, then 2 is definable as a successor of 1, and so on. Every natural number is definable. For every natural number there is a first-order property that identifies that number uniquely in $(\mathbb{N}, <)$. It implies that if a and b are different natural numbers then the types of a and b in $(\mathbb{N}, <)$ are different. Every number has its own unique type. Also, since every symmetry of a structure maps elements to elements of the same type, it means that $(\mathbb{N}, <)$ has no symmetries, other than the identity function. For each structure the identity function $f(x) = x$ on the domain is a symmetry. We call it trivial. A structure whose only symmetry is the trivial symmetry is called *rigid*. We have shown that $(\mathbb{N}, <)$ is rigid.

---

[2]A reminder: when we just say *definable*, we mean definable without parameters in the original language of the structure. If parameters are involved we say *parametrically definable*.

9.4 The Ordering of the Integers

Let us now compare $(\mathbb{N}, <)$ with the ordered set of the integers $(\mathbb{Z}, <)$. While every natural number is definable in $(\mathbb{N}, <)$, no integer is definable in $(\mathbb{Z}, <)$. Let us see why. Suppose a number a has a certain first-order property $\varphi(x)$. Let b be any other number. Let the function $f : \mathbb{Z} \longrightarrow \mathbb{Z}$ be defined by $f(x) = x + b - a$. If a is less than b, f shifts all numbers up by $b - a$, otherwise, it shifts them down by $a - b$. Since f preserves the ordering, it is a symmetry of $(\mathbb{Z}, <)$. Also $f(a) = a+b-a = b$. Hence, by Theorem 9.1, b also has the property $\varphi(x)$, and this proves that the set defined by $\varphi(x)$ in $(\mathbb{Z}, <)$ contains all integers.

The same argument shows that in $(\mathbb{Z}, <)$ there is only one type of single elements. We have shown, that if a and b are integers, then there is a symmetry f of $(\mathbb{Z}, <)$ such that $f(a) = b$; hence a and b have the same type in $(\mathbb{Z}, <)$. You can see that types of elements very much depend on the ambient structure. Each natural number has its own distinct type in $(\mathbb{N}, <)$, but they all share the same type in $(\mathbb{Z}, <)$.

There is only one type of single elements in $(\mathbb{Z}, <)$, but there are many types of ordered pairs. First let us observe that in any ordered structure, if $a < b$, then the types of (a, b) and (b, a) are different. The formula $x_1 < x_2$ is in the type of (a, b), and its negation is in the type of (b, a). Because of this, to avoid notational complications, we will only consider types of ordered pairs and sequences that are ordered from the lowest to the highest with respect to the ordering of the structure. In other words, we will talk about types of sequences $(a_1, a_2, \ldots, a_n)$ such that $a_1 < a_2 < \cdots < a_n$. As we will see, in some ordered sets, every two such ordered sequences of the same length have the same type. Not in $(\mathbb{Z}, <)$ though. If a and b are integers, and $a < b$, then let the *distance* between a and b be $b - a$. Let n be a natural number. The property "the distance between a and b is n" is definable in $(\mathbb{Z}, <)$. There is a specific definition for each n, and those definitions grow longer as n gets larger, but they all follow the same pattern. Here is a defining formula for $n = 3$:

$$\exists v \exists w[(x < v) \wedge (v < w) \wedge (w < y) \wedge \forall z[(x < z) \wedge (z < y) \Longrightarrow ((z{=}v) \vee (z{=}w))]].$$

If a and b are integers, and $a < b$, than the type of (a, b) is completely determined by the distance between a and b. That is because if (a, b) and (c, d) are two ordered pairs such that $a < b$, $c < d$, and the distances between a and b, and c and d are equal, then the symmetry that maps a onto c, must also map b onto d, which shows that the types (a, b) and (c, d) are the same. A similar argument applies to arbitrary finite ordered sequences. The type of an ordered increasing sequence $a_1, a_2, \ldots, a_n$ is completely determined by the distances between consecutive elements in the sequence.

$(\mathbb{Z}, <)$ is not minimal. For example, the set of negative numbers is defined by $x < 0$, and it is neither finite nor cofinite. There are other examples of definable sets, but not that many; $(\mathbb{Z}, <)$ is almost minimal. This last statement has a precise meaning, and to make it precise we need a couple of definitions.

Let $(A, <)$ be a linearly ordered set. For all a and b in A, the formula $(a < x) \wedge (x < b)$ defines the interval of all numbers between a and b. We call it the *open interval* between a and b. If a is not less than b, then the set defined by $(a < x) \wedge (x < b)$ is empty. We consider the empty set an open interval. The set A and sets of the form $\{x : x < a\}$ and $\{x : b < x\}$ are also open intervals of $(A, <)$. All open intervals of $(A, <)$ are parametrically definable and, so are the intervals of one of the forms

$$\{x : [(a < x) \wedge (x < b)] \vee (x = a)\},$$

$$\{x : [(a < x) \wedge (x < b)] \vee (x = b)\},$$

$$\{x : [(a < x) \wedge (x < b)] \vee (x = a) \vee (x = b)\}.$$

We call all such sets *intervals* of $(A, <)$.

For a, b, c and d in A, the formula $[(a < x) \wedge (x < b)] \vee [(c < x) \wedge (x < d)]$ defines the union of two open intervals. In a similar fashion one can write definitions for every finite union of intervals, including those that are unbounded at one end. For every finite set of intervals in a linearly ordered set, its is parametrically definable.

Definition 9.4 An *ordered structure* is a structure that includes a relation linearly ordering its domain. In particular, linearly ordered sets are ordered structures. An ordered structure is *order-minimal* if every subset of its domain that is parametrically definable is the union of finitely many intervals.[3]

One can show that $(\mathbb{Z}, <)$ is order-minimal, and the proof is similar to the one given in Chap. 12 that shows that $(\mathbb{N}, <)$ is minimal.

There are general theories of minimal and order-minimal structures, with important applications in classical algebra and analysis. Parametrically definable subsets of the domain of minimal or order-minimal structure are as simple as definable sets can be, but this is not implying anything directly about multi-dimensional sets that are definable in higher Cartesian powers of the domain. Those sets are more complex, but the theory provides precise insights into what this complexity is.

9.5 The Additive Structure of the Integers

Let us take a quick look and the additive structure of the integers. To fully describe the types and the subsets of $\mathbb{Z}$ that are parametrically definable in $(\mathbb{Z}, +)$ would take us too far into mathematics proper. Instead, we will make two important observations. The first is that the ordering is not definable in $(\mathbb{Z}, +)$, and the second is that $(\mathbb{Z}, +)$ is not minimal.

[3] In model theory, order-minimal structures are called *o-minimal*.

To prove that the ordering $<$ is not definable in $(\mathbb{Z}, +)$, we will use Theorem 9.1. Let f be the function that maps each element to its opposite: $f(x) = -x$. For all integers a, b, and c, if $a+b = c$, then by multiplying both sides by (-1) we get that $(-a) = (-b) + (-c)$. Hence, f is a symmetry of $(\mathbb{Z}, +)$; it preserves the addition relation, but does not preserve the ordering. It reverses it. If a formula $\varphi(x, y)$ in the first-order language of $(\mathbb{Z}, +)$ defined the ordering, then $\varphi(1, 2)$ would hold in $(\mathbb{Z}, +)$, and because f is a symmetry, $\varphi(-1, -2)$ would hold as well, which is a contradiction, showing that there can be no such definition.

The formula $\exists y[y+y = x]$ defines the set of even numbers and that set is neither finite nor cofinite in $\mathbb{Z}$. Hence, $(\mathbb{Z}, +)$ is not minimal. $(\mathbb{Z}, +)$ is not an ordered structure so we cannot ask whether it is order-minimal, but $(\mathbb{Z}, +, <)$ is ordered, and the same argument shows that this structure is not order-minimal.

One could ask whether addition is definable $(\mathbb{Z}, <)$. It is a legitimate question, especially since addition is in a strict sense determined by the ordering. For every integer a, $a + 1$ is the successor of a, and, as we have seen, the successor of a is definable in $(\mathbb{Z}, <)$ by a formula with parameter a. Similarly, for positive b, $a + b$ can be defined as the b-th successor of a. Let us see how it is done for $b = 3$. Let $s(x, y)$ be the formula of $(\mathbb{Z}, <)$ defining the relation "y is a successor of x." Then the following formula defines the relation $x + 3 = y$:

$$\exists v \exists w[s(x, v) \wedge s(v, w) \wedge s(w, y)].$$

For every natural number n we can define the relation $x + n = y$, but there is no single formula of $(\mathbb{Z}, <)$ that would define the relation $0 < y \wedge (x + y = z)$, even if parameters are allowed. This statement follows from order-minimality of $(\mathbb{Z}, <)$, but we can directly prove a weaker statement, namely that the relation $x + y = z$ is not definable without parameters in $(\mathbb{Z}, <)$. To this end, suppose $\varphi(x, y, z)$ defines addition in $(\mathbb{Z}, <)$. Consider $f : \mathbb{Z} \longrightarrow \mathbb{Z}$ defined by $f(x) = x + 1$. Since f is a symmetry of $(\mathbb{Z}, <)$, $\varphi(0, 0, 0)$ holds in $(\mathbb{Z}, <)$, and $f(0) = 1$; hence $\varphi(1, 1, 1)$ must also hold, telling us that 1+1=1, which is a contradiction.

The above argument cannot be used to prove that addition is not parametrically definable in $(\mathbb{Z}, <)$, because that would require symmetries that fix finite sequences of parameters, but there are no such symmetries. As soon as one integer is fixed, the whole structure becomes rigid. For any a in $\mathbb{Z}$, every integer is definable in $(\mathbb{Z}, <, a)$, either as an iterated successor or predecessor of a.

9.6 The Ordering of the Rational Numbers

In $(\mathbb{N}, <)$ every element has its own unique type. This implies that $(\mathbb{N}, <)$ is rigid. It has no nontrivial symmetries. The ordering of $\mathbb{Z}$ has symmetries, but not many. Every symmetry of $(\mathbb{Z}, <)$ is a shift moving all elements up, or all elements down. In other words, every symmetry of $(\mathbb{Z}, <)$ is of the form $f(x) = x + n$, for some integer n. The order structure of $\mathbb{Z}$ becomes rigid once we fix one of the elements, since then no shift can be applied.

The dense linear ordering $(\mathbb{Q}, <)$ has lots and lots of symmetries. There is no symmetry of $(\mathbb{Z}, <)$ that would fix 0 and move 1 to 2, but $f : \mathbb{Q} \longrightarrow \mathbb{Q}$ defined by $f(x) = 2x$ is a symmetry of $(\mathbb{Q}, <)$ and does it. $(\mathbb{Q}, <)$ is much more elastic than $(\mathbb{Z}, <)$. It is not hard to show that if $p_1, p_2, \ldots, p_n$ and $q_1, q_2, \ldots, q_n$ are increasing sequences of rational numbers, then there is a symmetry f such that $f(p_1) = q_1$, $f(p_2) = q_2, \ldots, f(p_n) = q_n$. It follows that not only all single rational numbers share the same type, but for each n, all increasing sequence of n rational numbers share the same type. Each pair (p_1, q_1) has the same type as any other pair (p_2, q_2) as long as they are ordered the same way. The notion of distance that was used to show that there are many types of ordered pairs in $(\mathbb{Z}, <)$ is meaningless in $(\mathbb{Q}, <)$.[4]

Using symmetries, it is not difficult to show that the structure $(\mathbb{Q}, <)$ is order-minimal. A more advanced reader can try to prove it as an exercise. Hints are provided in the exercise section. It is a bit more difficult to show that the ordered additive structure $(\mathbb{Q}, +, <)$ is also order-minimal. We will not go over details of the proof, but let us just see why the formula $\exists y[y + y = x]$ that defines even numbers in $(\mathbb{Z}, +, <)$ does not do that in $(\mathbb{Q}, +, <)$. The simple reason is that while in $\mathbb{Z}$ only even numbers are divisible by 2, in $\mathbb{Q}$ all numbers are. This is what the rational numbers were made for. A fraction of a fraction is again a fraction. By extending the number system from $\mathbb{Z}$ to $\mathbb{Q}$, we made the domain larger, but less complex because some specific properties, such as evenness, lost their meaning. Every rational number is even.

Something interesting is happening here. While the set of rational numbers $\mathbb{Q}$ is richer than the set of integers $\mathbb{Z}$, and the discrete ordering of $\mathbb{Z}$ seems less complex the dense ordering of $\mathbb{Q}$, nevertheless, the logical structure of $(\mathbb{Q}, +, <)$ is simpler than that of $(\mathbb{Z}, +, <)$. This pattern continues. We will see that $(\mathbb{R}, +, \cdot)$ is simpler than $(\mathbb{Q}, +, \cdot)$, and at the end that the field of complex numbers $(\mathbb{C}, +, \cdot)$, despite its name, is the simplest of them all—it is in fact minimal. To talk about all of that we need to say more about the geometry of definable sets in all those structures. We will do this in the next chapter.

Exercises

Exercise 9.1 *Show that the only subsets of the domain of a trivial structure that are definable without parameters are the empty set and the whole domain. Hint: Suppose that a and b are elements of the domain, and $\varphi(a)$ holds in the structure. Define a symmetry f such that $f(a) = b$.*

Exercise 9.2 *Show that every integer is definable in $(\mathbb{Z}, <, 0)$. Hint: See Exercise* 8.4.

[4]This should not be confused with the natural notion of distance for rational numbers. It is definable in $(\mathbb{Q}, +, <)$, but not in $(\mathbb{Q}, <)$.

Exercise 9.3 * *Prove that* $(\mathbb{Q}, <)$ *is order-minimal. Hint: Suppose that* $p_1, p_2 \ldots, p_n$ *is an increasing sequence of rational numbers, and the set* X *defined by a formula* $\varphi(x, p_1, p_2, \ldots, p_n)$ *in* $(\mathbb{Q}, <)$ *is nonempty. By defining an appropriate symmetry of* $(\mathbb{Q}, <, p_1, p_2, \ldots, p_n)$ *show that if* X *has a nonempty intersection with one of the intervals* $\{x : x < p_1\}$, $\{x : p_n < x\}$, *and* $\{x : p_i < x < p_{i+1}\}$, *for* $i = 1, \ldots, n-1$, *then* X *contains that whole interval.*

Chapter 10
Geometry of Definable Sets

Abstract In the previous chapter we saw examples of mathematical structures that are simple enough to allow a complete analysis of the parametrically definable subsets of their domains. Those structures are of some interest, but the real objects of study in mathematics are richer structures such as $(\mathbb{Q}, +, \cdot)$ or $(\mathbb{R}, +, \cdot)$. To talk about them we first need to take a closer look into their definable sets. Definable sets in each structure form a geometry in which the operations on sets are unions, intersections, complements, Cartesian products, and projections from higher to lower dimensions. We will see how those operations correspond in a natural way to Boolean connectives and quantifiers, and how the name "geometry" is justified when it is applied to sets definable in the field of real numbers. The last two sections are devoted to a discussion of the negative solution to Hilbert's 10th problem.

Keywords Boolean combinations · Existential quantification and projections · Diophantine equations · Hilbert's 10th problem · Tarski-Seidenberg theorem

10.1 Boolean Combinations

The geometry of definable sets involves two levels. At the first level, Boolean connectives[1] $\wedge$, $\vee$, and $\neg$ are used to combine the basic relations of the structure giving rise to new definable sets. The Boolean connectives correspond closely to basic operations on sets, which are also called *Boolean operations*. Conjunctions of formulas correspond to intersections of sets defined by them, disjunctions correspond to unions, and negations to complements. If A is the domain of a structure $\mathfrak{A}$, then for each natural number n, the definable subsets of A^n form a *Boolean algebra* of sets. Let $\mathfrak{B}$ be set of subsets of some set U. Then $\mathfrak{B}$ is a Boolean algebra, if for all X and Y in $\mathfrak{B}$ their intersection $X \cap Y$, their union $X \cup Y$, and their complements in U are also in $\mathfrak{B}$.

[1] After George Boole (1815–1864).

R. Kossak, *Mathematical Logic*, Springer Graduate Texts in Philosophy 3,
https://doi.org/10.1007/978-3-319-97298-5_10

At the next level, quantifiers are employed to bind free variables in formulas. A formula with n free variables defines a subset of the n-th Cartesian power of the domain. Appending quantifiers results in a formula with fewer free variables that defines a set in a lower Cartesian power. This is an important aspect of the geometry of definable sets and we will discuss it in detail in the next section.

Let $\mathfrak{A}$ be a structure with the domain A and two unary[2] relations R and S. We the use the same characters for relation symbols and the corresponding relations. Combining atomic formulas $R(x)$ and $S(x)$ using connectives $\wedge$, $\vee$, and $\neg$, we can form their *Boolean combinations*, such as $R(x) \wedge S(x)$, or $\neg R(x) \vee S(x)$. In the case of just two relation symbols, there is a helpful diagram that illustrates all possible combinations.

The picture below is the *Venn diagram* for the two sets R and S. The rectangle represents the domain of the structure. The two circles represent the sets R and S, their intersection represents the set defined by the formula $R(x) \wedge S(x)$, and the area outside the two circles is defined by $\neg R(x) \wedge \neg S(x)$. All Boolean combinations of $R(x)$ and $S(x)$ can be represented as regions in the diagram.

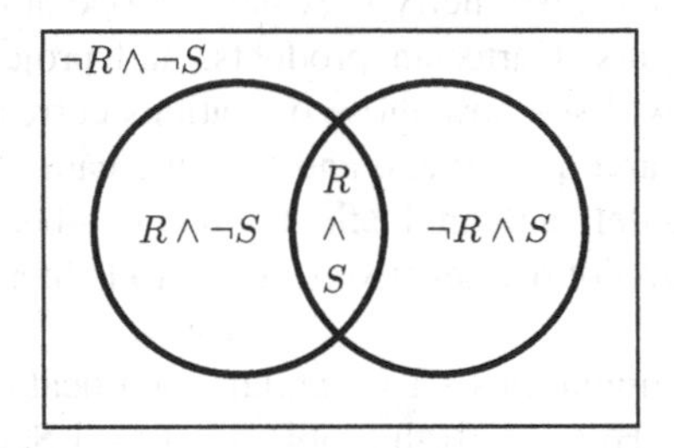

How many different Boolean combinations are there? In principle, there are infinitely many, because we can keep combining formulas to obtain longer and longer Boolean combinations, but by inspecting the diagram one can see that the only sets that can be defined by those formulas are combinations of different regions demarcated by the two circles. Cutting the rectangle along the circles will give us four pieces of the diagram. Any collection of those pieces corresponds to a subset of the domain—a unary relation that is definable from R and S. Since any choice of the pieces gives us a definable set, assuming that all pieces represent nonempty sets, altogether it gives us $2^4 = 16$ different definable sets, and that already includes the sets R and S, the empty set (no pieces), and the whole domain (all pieces).

To analyze all possible Boolean combinations one can obtain from three relations, we can inspect a Venn Diagram with three circles. Let us call the relations R, S, and T.

[2]Recall that a unary relation is a subset of the domain of a structure.

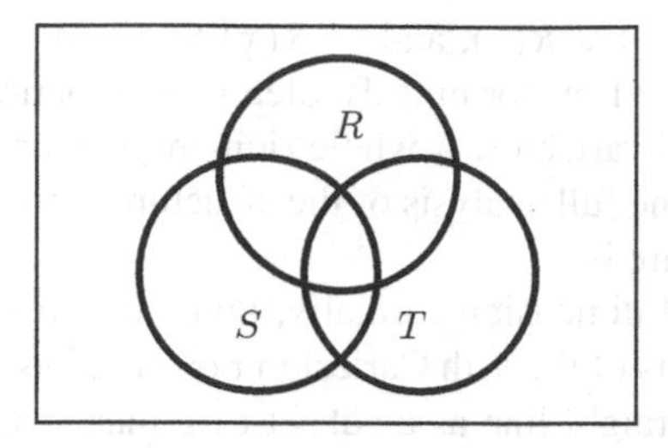

The three circles cut the domain into eight pieces. Any collection of those pieces gives us a definable set, so, again assuming they each piece represents a nonempty set, there are $2^8 = 256$ such definable sets. The diagram provides a visualization, and, in a sense, a complete understanding of what those sets are.

For more than three unary relations, no diagram with circles can represent all Boolean combinations, but there is nothing special about circles. One can use other geometric shapes, but the pictures are less transparent. The Wikipedia article on Venn diagrams gives interesting examples. Regardless of visual representations, the diagrams show that in a structure with a finite number of relations, Boolean combinations of the atomic formulas generate a large, but finite set of new definable relations.

In this introduction to the geometry of definable sets, so far we have only considered unary relations. In general, relations can be subsets of any Cartesian power of the domain of the structure. Boolean combinations of n-dimensional relations are definable subsets of the n-th Cartesian power of the domain. Why we call this geometry will become clearer in the next section in which we will see how logic allows us to define new sets in all possible dimensions, and how those sets interact with one another.

10.2 Higher Dimensions

Let us go back to the structure $\mathfrak{A}$ with domain A and two unary relations R and S. To define new relations other than the Boolean combinations of the basic relations, we will now take advantage of the infinite supply of variables. According to Definition 7.2, atomic formulas $R(x)$ and $R(y)$ define the same subset of A, namely the set R. Boolean combination $R(x) \wedge R(x)$, also defines R. Saying $R(x)$ twice does not make the statement any stronger. The situation changes when we consider $R(x) \wedge R(y)$. This formula has two free variables, hence it defines a subset of the Cartesian square A^2. It defines a set in two dimensions. Similarly, the formula $R(x) \wedge R(y) \wedge R(z)$ defines a subset of A^3, and in general the formula $R(x_1) \wedge R(x_2) \wedge \cdots \wedge R(x_n)$, with n free variables, defines a subset of A^n.

One could argue that the relation defined by $R(x) \wedge R(y)$ does not reveal any new features of the structure. Its "information content" is not much different than that of the set R itself. But a more complex picture emerges when we consider the relations such as $R(x) \vee R(y)$, and $\neg(R(x) \vee R(y))$, $R(x) \wedge S(y)$, and many other that can be obtained by forming Boolean combinations of atomic formulas with different choices of variables. A whole rich architecture of higher dimensional relations emerges, and the full analysis of the structure $\mathfrak{A}$ must include some insight into what this architecture is.

We have not defined dimension formally, and we will not do that. Informally, one could refer to subsets of the n-th Cartesian power A^n as n-dimensional sets, but this is not precise. A straight line as a subset of a plane, or the three-dimensional space, is still a one dimensional object. The n-th Cartesian power A^n has subsets of all dimensions up to n.

So far we have only considered definable sets that are Boolean combinations of the basic relations of the structure. It is time to take a look at the role of quantifiers. We will start with an example. In the field of real numbers $\mathfrak{R}$ the formula $4x^2 + y^2 - 16 = 0$ defines a subset E of $\mathbb{R}^2$ illustrated by the picture below.

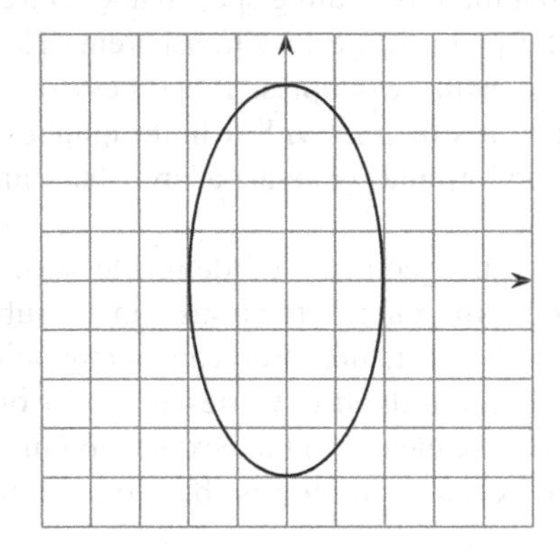

The set E is an ellipse with the minor axis of length 4 and the major axis of length 8. Let $\varphi(x)$ be the formula $\exists y[4x^2 + y^2 - 16 = 0]$, and let $\psi(y)$ be the formula $\exists x[4x^2 + y^2 - 16 = 0]$. Each of these formulas defines a set of real numbers. The first defines the interval of all numbers between -2 and 2, including the endpoints. The second defines the interval between -4 and 4, also with the endpoints. Those two sets are images under *projections* of the set E onto the horizontal and the vertical axes. Binding free variables in a formula by existential quantifiers results in definitions of images under projections of definable sets in higher dimensions onto lower dimensions. Projections are among the most important geometric operations; hence, this feature of existential quantification provides a strong link between geometry and logic. Let us examine this phenomenon in greater detail.

A quadratic equation in two variables is an equation of the form

$$ax^2 + bxy + cy^2 + dx + ey + f = 0, \qquad (*)$$

where x and y are variables representing the unknowns, and a, b, c, d, e, and f represent parameters. In our example above $a = 4$, $c = 1$, $f = -16$, and all other parameters are equal to 0. Each equation of the form (∗), in which the parameters are real numbers, is a formula of the language of $\overline{\mathfrak{R}}$, and it defines a subset of $\mathbb{R}^2$. Such sets are known as *conic sections*. Each conic section is either a circle, an ellipse, a parabola, a hyperbola, or, when both a and c are 0, a straight line, or two straight lines, as, for example, is the case for the equation $x^2 - y^2 = 0$.[3] For some choice of parameters, the corresponding conic section is empty. For example, it is empty for the equation $x^2 + y^2 + 1 = 0$ which has no real solutions.

Since the order relation $<$ is definable in $\mathfrak{R}$, the solution sets to inequalities of the form

$$ax^2 + bxy + cy^2 + dx + ey + f < 0, \qquad (**)$$

are also definable in $\overline{\mathfrak{R}}$. The formula $4x^2 + y^2 - 16 < 0$ defines the interior of the ellipse E.

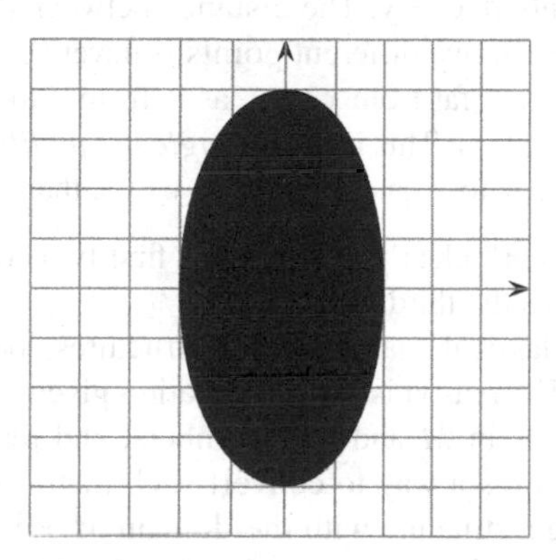

Boolean combinations of the solution sets to equations (∗) and inequalities (∗∗) are also definable in $\overline{\mathfrak{R}}$. We already see some interesting geometry, and it is only based on polynomial equations in two variables and of degree 2. A much richer picture emerges when we also consider polynomials of higher degrees, and equations and inequalities with more variables. It becomes even more interesting when we also include projections from higher to lower dimension, but before we see how, we need a few more words of explanation why we call all this a geometry.

10.2.1 Euclidean Spaces

The Cartesian powers $\mathbb{R}^n$ serve as models of classical geometry. $\mathbb{R}^1$ is just the one-dimensional line $\mathbb{R}$; $\mathbb{R}^2$ is the two-dimensional plane; $\mathbb{R}^3$ is the three-dimensional space; and so on. Points in $\mathbb{R}^2$ are ordered pairs of numbers, points in $\mathbb{R}^3$ are ordered triples, and in general a point in $\mathbb{R}^n$ is a sequence of n real numbers. Each space $\mathbb{R}^n$

[3]Notice that another special case $ax^2 + by^2 = 0$ the defined set is just $\{(0, 0)\}$. One point can be considered as a circle that has radius 0.

is equipped with a metric, that is a function computing distances between points. The distance between two points $(x_1, \ldots, x_n)$ and $(y_1, \ldots, y_n)$ in $\mathbb{R}^n$ is

$$\sqrt{(x_1 - y_1)^2 + \cdots + (x_n - y_n)^2}.$$

For example, the distance between $x_1 = 2$ and $y_1 = 5$ in $\mathbb{R}^1$ is $\sqrt{(2-5)^2} = \sqrt{9} = 3$.[4] The distance between $(x_1, x_2) = (0, 0)$ and $(y_1, y_2) = (1, 1)$ in $\mathbb{R}^2$ is $\sqrt{(1-0)^2 + (1-0)^2} = \sqrt{2}$.

The set $\mathbb{R}^n$ with the distance function defined above is known in mathematics as an *Euclidean metric space*. In general, a *metric space* is a set M equipped with a distance function d, defined on M^2 with values in $\mathbb{R}$, which satisfies the following conditions for all x, y, and z in M:

- $d(x, y) = 0$ if and only if $x = y$. The distance between the point and itself is 0, and the distance between two different points is never 0.
- $d(x, y) = d(y, x)$. It is as far from x to y, as is from y to x.
- $d(x, z) \leq d(x, y) + d(y, z)$. This is the *triangle inequality*. It says that it is never shorter to go first from x to y, and then from y to z, than directly from x to z.

The distance formula for $\mathbb{R}^n$ clearly satisfies the first two conditions, and it can be shown that it also satisfies the third.

Metric spaces are classical mathematical structures, but they do not fit our definition of structure. The reason is that the relation given by the distance function is between pairs of points in M and real numbers, and real numbers are usually not elements of M. There is a way to convert each metric space into a first-order structure by considering a structure with the domain $M \times \mathbb{R}$, but in the case of the Euclidean spaces $\mathbb{R}^n$, $\mathbb{R}^n \times \mathbb{R}$ can be identified with $\mathbb{R}^{n+1}$, so there is no need to do that. The distance function is a relation between pairs of points $(x_1, \ldots, x_n)$ and $(y_1, \ldots, y_n)$ and a number r that is the distance between them. The points $(x_1, \ldots, x_n)$, $(y_1, \ldots, y_n)$, and the number r are related when

$$0 \leq r \wedge (x_1 - y_1)^2 + \cdots + (x_n - y_n)^2 = r^2.$$

The distance function for $\mathbb{R}^n$ can be identified with a definable relation on the set $\mathbb{R}^{2n+1}$. This makes all Euclidean geometry a part of the first-order structure of $\overline{\mathfrak{R}}$.

Both projections of the ellipse E in our example above are intervals. It turns out that if a subset X of $\mathbb{R}^n$ is defined by a first-order formula $\varphi(x_1, x_2, \ldots, x_n)$ of the language of $\overline{\mathfrak{R}}$, then the projection of X onto $\mathbb{R}$, defined by $\exists x_2 \cdots \exists x_n \varphi(x_1, x_2, \ldots, x_n)$, is a finite union of intervals. It is not difficult to

[4] The distance between two numbers a and b in $\mathbb{R}$ is defined to be $\sqrt{(a-b)^2}$, which can also be defined more naturally as $|a - b|$.

see that this is the case of Boolean combinations of conic sections. In general, it is a theorem proved independently by Abraham Seidenberg and Alfred Tarski. The crucial part of the proof is to show that if a subset X of R^n, with $n > 1$, is definable in $\overline{\mathfrak{R}}$ by a formula with no quantifiers, then its projection obtained by appending an existential quantifier in front of the formula, is again definable by a formula without quantifiers. It follows that in $\overline{\mathfrak{R}}$ every formula is equivalent to a formula without quantifiers. In the analysis of the geometry of sets that are definable in $\overline{\mathfrak{R}}$ one only has to consider sets definable by such formulas. The quantifiers have been eliminated. This is a actually a technical term: we say that $\overline{\mathfrak{R}}$ admits *elimination of quantifiers*.

Our statement of the Tarski-Seidenberg theorem is not quite correct. Everything is fine if we assume that the language of $\overline{\mathfrak{R}}$ includes the relation symbol $<$. Previously we have noted that one does not need to extend the language of $\overline{\mathfrak{R}}$ to include $<$. Addition and multiplication suffice because the ordering relation $<$ is definable in $\mathfrak{R}$. Now we have to be more careful. The defining formula is $a < b$ iff $\exists z[\neg(z = 0) \wedge (a + z \cdot z = b)]$. It has an existential quantifier. That quantifier cannot be eliminated. It can be shown that every definition of the ordering of $\mathbb{R}$ must use at least one quantifier. The Tarski-Seidenberg elimination of quantifiers for $\overline{\mathfrak{R}}$ uses the ordering in an essential way.

The discussion above focused on existential quantifiers. What about the universal ones? Universal quantifiers are indispensable in expressing various properties of structures in a natural way, but they can always be eliminated: instead of $\forall x \varphi(x)$ we can say: $\neg\exists x \neg\varphi(x)$. This shows that if all existential quantifiers can be eliminated in definitions of sets over a structure, then universal quantifiers can be eliminated as well.

10.3 Shadows and Complexity

Intuitively, the image under projection of a set from a higher dimension to lower should be less complex than the set itself. After all, all features of the shadow derive from the object, and some particular features may get lost. This is an example of how geometric intuition can fail us. Henri Lebesgue was one of the two most prominent mathematicians of the first quarter of the twentieth century.[5] The history of mathematics records a famous Lebesgue error concerning Borel sets.[6] Borel sets

[5]The other one was David Hilbert.

[6]If X is a metric, or in general a topological space, then the set of all Borel sets of X is the smallest set that contains all open subsets of X and is closed under complements, and countable unions and intersections i.e. if A is Borel, then so is $X \setminus A$, and if $A_1, A_2 \ldots$ is a sequence of Borel sets, then the union and the intersection of all sets in that sequence are also Borel.

can be quite complex, but from a certain point of view they are considered relatively simple. Lebesgue thought that the image under projection of a Borel subset of a plane $\mathbb{R}^2$ onto any of the coordinate axes is also Borel. So Lebesgue thought that the shadow of a relatively simple set is relatively simple, and he even gave a (false) proof of it. Lebesgue's error was found and corrected by Suslin, who showed that the image under projection of a relatively simple set in $\mathbb{R}^2$ can be much more complex than the projected set.

In the logic approach, we measure complexity of a definable set by the smallest number of quantifiers needed for its definition. Sets that can be defined without quantifiers have complexity 0, sets that can be defined by a formula with only one quantifier have complexity 1, and so on. In the case of the field of real numbers in the language including the relation symbol $<$, by the theorem of Tarski and Seidenberg all quantifiers can be eliminated, hence all definable sets have complexity 0. In a sharp contrast, in seemingly simpler structures $(\mathbb{N}, +\cdot)$, $(\mathbb{Z}, +\cdot)$, $(\mathbb{Q}, +\cdot)$, for each n there are sets that can be defined by formulas with $n+1$ quantifiers, but cannot be defined by any formula with n quantifiers. A full explanation would require mathematical tools beyond the scope of this book, but in the following subsection we will discuss examples that may shed some light on where the complexity of those three structures comes from.

Let $\overline{\mathfrak{Z}}$ be the structure $(\mathbb{Z}, +, \cdot)$ expanded by adding names for all integers. All integers are definable $(\mathbb{Z}, +, \cdot)$, which shows that every relation definable in $\overline{\mathfrak{Z}}$ is already definable in $(\mathbb{Z}, +, \cdot)$, but since eliminating parameters in formulas by replacing them by their definitions adds additional quantifiers, and since we are paying close attention to quantifier complexity of formulas, now we will throw all the parameters into the language.

A bit surprisingly, $\overline{\mathfrak{Z}}$ turns out to be much more complex than $(\mathbb{R}, +, \cdot)$. In fact, $\overline{\mathfrak{Z}}$ is one of the most complex structures in mathematics.

To understand better the peculiar structure of $\overline{\mathfrak{Z}}$, and how is differs from $\overline{\mathfrak{R}}$ we need to discuss solvability of polynomial equations. We will do it in the next section by examining some prominent examples.

10.3.1 Diophantine Equations and Hilbert's 10th Problem

Let us begin with the equation $x^2 + y^2 = 1$. This equation has infinitely many real solutions. The solutions form a circle in $\mathbb{R}^2$ that is centered at $(0, 0)$ and has radius 1. Only four solutions have integer coordinates. They are (0,1), (1,0), (0,−1), and (−1,0). By analogy, one is tempted to say that those four points *are* a circle in $\mathbb{Z}^2$.

We will focus on the question whether a given polynomial equation has solutions in $\mathbb{Z}$, or, in other words, whether the set defined by the equation in $\overline{\mathfrak{Z}}$ is nonempty.

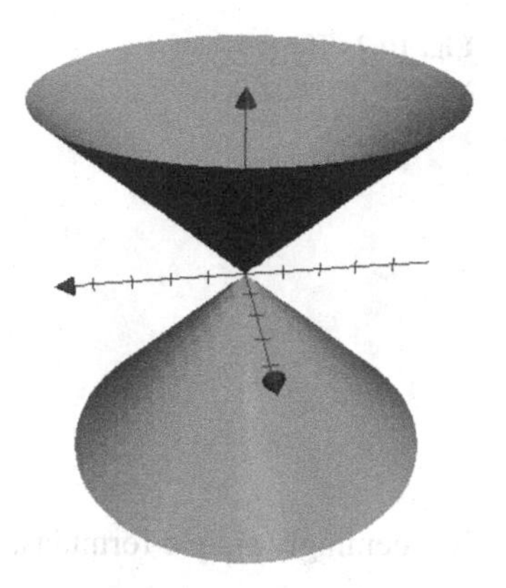

Fig. 10.1 The graph of $x^2 + y^2 = z^2$

The equation $x^2 + y^2 = 2$ also has four solutions in $\mathbb{Z}^2$: (1,1), $(-1, -1)$, $(-1, 1)$, and $(1, -1)$, but $x^2 + y^2 = 3$ has none, which can be quickly verified by direct checking, since the only integer candidates for x and y are 1, -1, and 0. The the last equation defines the circle in $\mathbb{R}^2$ that is centered at (0,0) and has radius $\sqrt{3}$, but each point (x, y) of that circle has at least one non-integer coordinate.

Let us now consider $x^2 + y^2 = z^2$. In $\mathfrak{R}$, this equation defines a subset of R^3. The solution set consists of two symmetric cones (Fig. 10.1). Some of the points in the solution set are in $\mathbb{Z}^3$, for example (0, 0, 0), or (3, 4, 5). In fact, the equation has infinitely many integer solutions. For every integer k, $(3k, 4k, 5k)$ is in the solution set, but there are many other.

Integer solutions of the equation $x^2 + y^2 = z^2$ are called *Pythagorean triples*. They are all known and classified, but this classification involves a good deal of elementary number theory, and it shows that the set defined by the equation in $\overline{\mathfrak{Z}}$ is quite complex.

The next example is $x^3 + y^3 = z^3$. This equation has trivial solutions, such as (0, 0, 0) or $(-2, 2, 0)$, but it has no solutions in which x, y, and z are positive integers (Fig. 10.2).

While there are many squares that are sums of two squares, the sum of two positive cubes can never be a cube.[7] In fact, for any n greater than 2, the equation $x^n + y^n = z^n$ has no positive integer solutions. This was observed by Pierre de Fermat (1601 or 1607–1665), the French lawyer and mathematician who on the margin of Diophantus' *Arithmetica* famously wrote "It is impossible to separate a cube into two cubes, or a fourth power into two fourth powers, or in general, any power higher than the second, into two like powers. I have discovered a truly marvelous proof of this, which this margin is too narrow to contain." This statement, called Fermat's Last Theorem, was finally proved in 1994 by Andrew Wiles. The proof is immensely difficult, and it relies on a whole range of deep results from several areas of mathematics. For us, it offers a caveat: deciding whether sets defined

[7] A square is a number of the form n^2, and a cube is a number of the form n^3.

Fig. 10.2 The graph of $x^3 + y^3 = z^3$

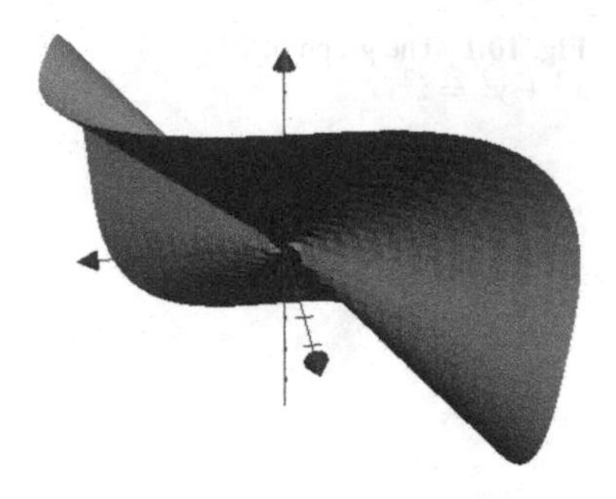

by seemingly simple formulas, such as

$$x > 0 \wedge y > 0 \wedge z > 0 \wedge x^3 + y^3 = z^3,$$

are nonempty may require a lot of hard work.[8]

What makes such a big difference between $x^2 + y^2 = z^2$ and $x^n + y^n = z^n$, for $n > 2$? The difference is striking, but mathematical reasons for it are not easy to sort out. The syntax of the two formulas is almost identical, but the sets that the formulas define in $\mathfrak{Z}$ are very different. It is hard to know what the solution sets are just by just analyzing the form of their definitions.

There is a whole area of number theory, known as the theory of Diophantine equations,[9] that is devoted to the study of integer solutions of polynomial equations. Much of it consists of a painstaking analysis of particular cases. In this context, one should appreciate the audacity of David Hilbert who, addressing the International Congress of Mathematicians in Paris in 1900, included among the 23 challenge problems for the twentieth century mathematics his problem number 10: "Given a Diophantine equation with any number of unknown quantities and with rational integral numerical coefficients: To devise a process according to which it can be determined in a finite number of operations whether the equation is solvable in rational integers." In other words, Hilbert was asking if there is a mechanical procedure to decide whether a polynomial equation of the kind we are discussing has integer solutions. The meaning of the question at the time of Hilbert was equivocal, since the notion of a process that can be used to decide a mathematical question in a finite number of steps, while intuitively clear, had not been formalized yet.

Hilbert could not know that analogous question for the field of real numbers $\mathfrak{R}$ has a positive solution—the Tarski-Seidenberg theorem was proved only in the 1940s—but certain partial decision procedures of the kind Hilbert was asking for

[8]It was known well before Wiles' proof that the equation $x^3 + y^3 = z^3$ has no positive integer solutions. This was proved by another great mathematician Leonard Euler (1707–1783). Many other cases of Fermat's Last Theorem had been proved before Wiles announced his result. The smallest exponent n for which the theorem had not been verified before, is 4,000,037. This last fact is not very well-known. I am grateful to my colleague Cormac O'Sullivan for digging it up.

[9]After the third century Greek mathematician Diophantus of Alexandria.

had been known for a long time. To give a high school example, to find out if the equation $ax^2 + bx + c = 0$, where a, b, and c are integers, has a solution in $\mathbb{R}$, one can compute the number $\Delta = b^2 - 4ac$. The equation has a solution in $\mathbb{R}$ if and only if $\Delta \geq 0$, and has a solution in $\mathbb{Q}$ if and only if Δ is a square. Hilbert was asking for something much more general. The mechanical procedure he was asking for would give the answer for any polynomial equation with integer coefficients and any number of variables, and it would tell us if there are solutions in $\mathbb{Z}$.

Fermat's Last Theorem had to wait for its proof roughly 360 years. Hilbert's 10th problem was solved 70 years after it was posed, and the solution was negative. The solution is the celebrated MRDP theorem, proved in a collaborative effort over many years by Martin Davis, Yuri Matiyasevich, Hilary Putnam, and Julia Robinson.[10]

To see what the MRDP theorem has to do with shadows of definable sets, let us look again at equations defining conic sections in $\mathbb{R}^2$. We have to go through some technicalities, and we need some more notation.

Recall that a conic section in $\mathbb{R}^2$ is the set of points (x, y) defined in $\overline{\mathfrak{R}}$ by the formula

$$ax^2 + bxy + cy^2 + dx + ey + f = 0, \qquad (*)$$

where the parameters a, b, c, d, e, and f can be any real numbers. The solution set could be empty. For example, when $a = b = f = 1$, and all other parameters equal 0, the resulting equation is $x^2 + y^2 + 1 = 0$. This equation has no solutions, as the value on the left-hand side for any x and y is always at least 1.

Now the notation will become a bit more complicated. If this were a mathematics textbook, we would promptly introduce more notation so that the formulas we want to examine would not look that daunting. This is a good practice, but it takes time to learn the abbreviations, and to understand what they hide, so we will not do it here. The formulas will look rather complex, but what we are going to do is rather straightforward. We will make a small change in $(*)$ by replacing all parameters by variables. The equation $(*)$ becomes

$$z_1x^2 + z_2xy + z_3y^2 + z_4x + z_5y + z_6 = 0. \qquad (**)$$

Further, let $p(z_1, z_2, z_3, z_4, z_5, z_6, x, y)$ be the polynomial $z_1x^2 + z_2xy + z_2y^2 + z_4x + z_5y + z_6$. We are splitting hairs here, the only difference between $(*)$ and $(**)$ is that in the former a, b, c, d, e, and f represent arbitrary but fixed parameters in the equation; hence $(*)$ is a general form of an equation with two unknowns x and y. It represents infinitely many such equations, one for each selection of parameters. In $(**)$, we replaced the parameters by variables, so now it is a single equation with eight unknowns.

[10] For a very touching personal account of the history of the MRDP theorem see [29]. Technical details are included.

Let S be the subset of $\mathbb{R}^6$ defined by the formula

$$\exists x \exists y[p(z_1, z_2, z_3, z_4, z_5, z_6, x, y) = 0].$$

The quantifier-free formula

$$p(z_1, z_2, z_3, z_4, z_5, z_6, x, y) = 0$$

defines a subset A of $\mathbb{R}^8$. The set S is the image of A under the projection onto $\mathbb{R}^6$. According to the Tarski-Seidenberg theorem, S has a quantifier-free definition, which means that there is a polynomial $q(z_1, z_2, z_3, z_4, z_5, z_6)$ such that S is the set of those $(z_1, z_2, z_3, z_4, z_5, z_6)$ in $\mathbb{R}^6$ for which $q(z_1, z_2, z_3, z_4, z_5, z_6) = 0$.

The Tarski-Seidenberg theorem also provides an algorithmic procedure that given the polynomial p as above, produces q with the required property. Now, for the given integer parameters a, b, c, d, e, and f, one can compute $q(a, b, c, d, e, f)$ and if the result is 0, this means that the equation $(*)$ with those numbers as parameters has a solution in $\mathbb{R}^2$, and otherwise it does not.

Compare the two definitions of the set S. The first one employs a definition with two existential quantifiers; the second is quantifier-free. Deciding if the equation $(*)$ has a solution using the first definition involves a potentially infinite search for x and y; hence, while formally well-defined, it may be of little use. Searching for suitable x and y in a vast infinite domain is, in general, not a feasible task. In contrast, the second definition involving the polynomial q is based on a straightforward computation.[11]

We have outlined a decision procedure for solvability of arbitrary quadratic equations in two unknowns x and y. The same procedure applies to all polynomial equations in any number of variables.

Let us now see what changes when instead of solvability of equations in $\mathbb{R}$ one is interested in solutions in $\mathbb{Z}$. One of the consequences of the MRDP theorem is that there is a polynomial

$$h(z_1, z_2, \ldots, z_m, x_1, x_2, \ldots, x_n),$$

in which the variables $z_1, z_2, \ldots, z_m$ represent parameters, and variables $x_1, x_2, \ldots, x_n$ represent unknowns, such that there is no algorithmic procedure to decide whether for given integer parameters $a_1, a_2, \ldots, a_m$, there are integers $k_1, k_2, \ldots, k_n$ such that

$$h(a_1, a_2, \ldots, a_m, k_1, k_2, \ldots, k_n) = 0. \qquad (\dagger)$$

[11] In practice it may not be as simple as it appears, and in fact quite often it is not. The polynomial $q(z_1, z_2, z_3, z_4, z_5, z_6)$ may be of high degree, and calculations needed to evaluate it may be tedious, or simply too hard, even for a fast computer.

Contrary to what Hilbert himself believed, his 10th problem not only has a negative solution in general; the solution is negative even in the case of parametric versions of a single polynomial.

Much advanced mathematics goes into details of the results in this section. In particular, the construction of the polynomial h is messy, and there is still research going on to minimize both the number of parameters in $z_1, z_2, \dots, z_m$, and the number of variables in $x_1, x_2, \dots, x_n$ in it.

Now we can go back to projections. Let S be the set of those parameters $a_1, a_2, \dots, a_m$, for which the equation (†) above has a solution $(k_1, k_2, \dots, k_n)$ in $\mathbb{Z}^n$. Then, S is a subset of $\mathbb{Z}^{m+n}$, and it is defined in $\overline{\mathfrak{Z}}$ by a formula with n existential quantifiers

$$\exists x_1 \exists x_2 \cdots \exists x_n [p(z_1, z_2, \dots, z_m, x_1, x_2, \dots, x_n) = 0].$$

We cannot replace this definition with an equivalent quantifier-free formula, because if we could, that would give us a computational procedure to check membership in the set S, but we know from the MRDP theorem that there is no such procedure. So this is our example. The set A of all $(z_1, z_2, \dots, z_m, x_1, x_2, \dots, x_n)$ such that

$$p(z_1, z_2, \dots, z_m, x_1, x_2, \dots, x_n) = 0$$

is simple, because a calculation involving only addition and multiplication of integers can reveal if a point in $\mathbb{Z}^{m+n}$ is in it or not. Its projection S is much more complex. A given sequence $(a_1, a_2, \dots, a_m)$ may be in A or may be not, but we do not have any way of checking by a computation.

So here is our conclusion: in $\overline{\mathfrak{Z}}$ the image under projections of a simply defined set can be much more complex then the set itself. The interesting part of this story is that this syntactic complexity of the set, as measured by the number of quantifiers in its definition does not determine the complexity of projected images. The final outcome depends on semantics; it depends on the structure in which the language is interpreted. First-order theories of $\overline{\mathfrak{Z}}$ and $\overline{\mathfrak{R}}$ share the same syntax, but their semantics are very different.

10.3.2 The Reals vs. The Rationals

One could object that the comparison between $\overline{\mathfrak{Z}}$ and $\overline{\mathfrak{R}}$ is unfair. Why should one expect the geometries of definable sets in those two structures to be similar? Just by looking at the ordered sets $(\mathbb{Z}, <)$ and $(\mathbb{R}, <)$ one sees stark differences. The former is a countable (small) discrete order, while the latter is uncountable (huge) and dense. With a few exceptions, each polynomial equation in two unknowns defines a smooth curve in $\mathbb{R}^2$, but only rarely do those curves pass through points both of whose coordinates are integers. It seems reasonable to expect that the geometry of $\mathbb{Z}^2$ is much more complex than that of $\mathbb{R}^2$, as indeed it is.

Algorithms for addition and multiplication of integers represented in decimal notation are straightforward, but the simplicity of these operations is deceptive. In polynomial equations, addition and multiplication are mixed together, and this creates complexity. As we have seen, this complexity disappears when we move to the much smoother structure $\overline{\mathfrak{R}}$. The transition from $\mathbb{Z}$ to $\mathbb{R}$ proceeds via the intermediate structure of rational numbers $\mathbb{Q}$, so now we need to discuss how complex are the definable sets in that structure.

Many equations with integer coefficients, such as $x + x = 1$, do not have integer solutions. The set defined in $\overline{\mathfrak{Z}}$ by $x + x = 1$ is empty, but in $(\mathbb{Q}, +)$ it defines the one element set $\{\frac{1}{2}\}$. Many more equations can be solved in $\mathbb{Q}$ than in $\mathbb{Z}$, but not all. For example $x \cdot x = 2$, has no solutions in $\mathbb{Q}$, but it has two solutions in $\mathbb{R}$. The set of rational numbers defined by this equation is empty, but in $\overline{\mathfrak{R}}$ it defines the two element set $\{-\sqrt{2}, \sqrt{2}\}$. In the evolution of the number system we moved from the simpler and easier to describe set $\mathbb{Q}$ to a much larger and somewhat mysterious $\mathbb{R}$ exactly for this reason—to be able to find all solutions to polynomial equations that represent geometric quantities. The domain $\mathbb{R}$ is incomparably more complex than the simple set $\mathbb{Z}$, but the additive and multiplicative structure of the real numbers is much smoother, and in effect the geometry of parametrically definable sets becomes much easier to describe.

The domain $\mathbb{Q}$ is countable, so not as large as $\mathbb{R}$, but the ordering is dense, and it is a field. Let $\overline{\mathfrak{Q}}$ be $(\mathbb{Q}, +, \cdot)$ in the language with names for all rational numbers. On the one hand, because of the density of its order, $\overline{\mathfrak{Q}}$ looks a bit like $\overline{\mathfrak{R}}$, on the other hand, because many polynomial equations do not have rational solutions, it inherits some complexity from $\overline{\mathfrak{Z}}$. In terms of logical complexity $\overline{\mathfrak{Q}}$ is somewhere between $\overline{\mathfrak{Z}}$ and $\overline{\mathfrak{R}}$, but it turns out that $\overline{\mathfrak{Q}}$ is nowhere near $\overline{\mathfrak{R}}$.

The geometry of definable relations on $\overline{\mathfrak{Q}}$ is as complex as that of $\overline{\mathfrak{Z}}$, but this is neither clear nor easy to prove. It follows from a theorem of Julia Robinson, proved in 1949. Robinson found a formula that defines the set $\mathbb{Z}$ in $\overline{\mathfrak{Q}}$. This showed that to be an integer is a first-order property in the field of rational numbers. The great complexity of $\overline{\mathfrak{Z}}$ is present in $\overline{\mathfrak{Q}}$. This statement may seem obvious, since $\mathbb{Z}$ is a subset of $\mathbb{Q}$, but the fact that $\mathbb{Z}$ is logically visible $\overline{\mathfrak{Q}}$ makes $\mathbb{Z}$ a part of the geometry of the field of rational numbers. Recall that, by the Tarski-Seidenberg theorem, $\mathbb{Z}$ is not parametrically definable in the field of real numbers.

The field of rational numbers is at least as complex as $\overline{\mathfrak{Z}}$, but it is not more complex than $\overline{\mathfrak{Z}}$. When we introduced the structure $(\mathbb{Q}, +, \cdot)$, we did it by defining it in a first-order way in $(\mathbb{Z}, +, \cdot)$, hence the former structure is logically no more complex than the latter: any relation that is logically visible in $(\mathbb{Q}, +, \cdot)$ is also visible in $(\mathbb{Z}, +, \cdot)$, and vice versa. From the point of view of logical analysis, $\overline{\mathfrak{Z}}$ and $\overline{\mathfrak{Q}}$ are equivalent.

Exercises

Exercise 10.1 *Show that every integer is definable in the structure* $(\mathbb{Z}, +, \cdot)$. *Write explicit definitions of* 3 *and* -3, *try to use as few quantifiers as possible. What is the quantifier complexity of your definition? Hint: See Exercise* 9.2.

Exercise 10.2 *By direct checking, find all integer solutions of the equation* $x^2 + y^2 = 4$, *and show that* $x^2 + y^2 = 3$ *has none.*

Exercise 10.3 * *Express the Pythagorean theorem by first-order sentence in the language with* $+$ *and* $\cdot$. *Hint: Let* $A = (a_1, a_2)$, $B = (b_1, b_2)$, *and* $C = (c_1, c_2)$ *be points in* $\mathbb{R}^2$. *Then the segments* BA *and* BC *form a right angle if and only if*

$$(a_1 - b_1) \cdot (c_1 - b_1) + (a_2 - b_2) \cdot (c_2 - b_2) = 0.$$

Exercise 10.4 * *Suppose that for every quantifier-free formula* $\varphi(x, x_1, \ldots, x_n)$ *of the language of a structure* $\mathfrak{A}$, *there is a quantifier-free formula* $\psi(x_1, \ldots, x_n)$ *such that for all* $a_1, \ldots a_n$ *in the domain of* $\mathfrak{A}$, $\exists x \varphi(x, a_1, \ldots, a_n)$ *holds in* $\mathfrak{A}$ *if and only if* $\psi(a_1, \ldots, a_n)$ *holds in* $\mathfrak{A}$. *Show that every sentence* φ *of the language of* $\mathfrak{A}$ *is equivalent to a quantifier-free sentence of that language. Hint: Use induction on the logical complexity of* φ.

Chapter 11
Where Do Structures Come From?

Abstract The compactness theorem, Theorem 11.2, is one of the most frequently used basic tools of model theory. It implies that for every structure with an infinite domain there is another structure that is very similar but not isomorphic to the given one. We will see a toy example that shows how such structure could be used to study number-theoretic problems. A more advanced application is given in Appendix A.5.

Keywords Completeness theorem · Compactness theorem · Elementary extensions · Twin primes conjecture · Nonstandard models

11.1 The Compactness Theorem

A mathematical structure is a set with a set of relations on it. The universe of sets is rich and diverse, hence so is the world of all structures. The axioms of set theory allow us to prove that there exist set-theoretic representations of all objects that modern mathematics needs. This is good, but one may become concerned that the axioms are too powerful. Do we really need all those structures that set theory provides? The answer is positive, but not straightforward. We have seen how set theory is used to construct representations of structures that we intend to study. In this chapter we will see how it can also be used to generate the whole range of other structures often with unexpected properties.

For a given structure, one can ask whether some familiar specific property is expressible in first-order logic, but now we will ask a more ambitious question: Given a structure $\mathfrak{A}$, can we describe *all* properties of $\mathfrak{A}$ that are first-order expressible? Put in this way, the problem is somewhat vague. To make it precise let us define the *first-order theory of* $\mathfrak{A}$ to be the set of all first-order sentences that are true in $\mathfrak{A}$. There are infinitely many sentences that are true in $\mathfrak{A}$, and infinitely many that are false. How can we possibly know the truth or falsity for all of them?

In some cases, the task is not that hopeless. It can happen that there is a finite set of sentences in the theory of a structure from which all other sentences that

R. Kossak, *Mathematical Logic*, Springer Graduate Texts in Philosophy 3,
https://doi.org/10.1007/978-3-319-97298-5_11

are true in the structure logically follow.[1] If this happens, we say that the theory of $\mathfrak{A}$ is *finitely axiomatizable*. There are structures whose theories are not finitely axiomatizable, but for which there is an infinite, effectively listed set of sentences from which all other first-order truths about the structure follow. In such cases, we say that the theory of the structure is *axiomatizable*.[2] If the theory of a structure is axiomatizable, it may not be a trivial matter to derive this or that particular true statement from the axioms, but at least we do know where all those true statements come from.

Examples of structures with axiomatizable theories are $(\mathbb{Q}, <)$, and $(\mathbb{Z}, +)$; the former is axiomatized by a single sentence expressing that the ordering of $\mathbb{Q}$ is dense and has no least and no last element; the latter by a set of sentences expressing elementary properties of addition. An example of a structure whose theory is not axiomatizable is $(\mathbb{Z}, +, \cdot)$.

Axiomatizability of theories of structures is a big topic and a fuller discussion would take us too far. We will move in another direction.

By a *first-order theory*, we simply mean a set of sentences in a first-order language. Let T be a theory. A structure in which all sentences in T are true is called a *model* of T.

For any structure $\mathfrak{A}$, the theory of $\mathfrak{A}$, denoted $\mathrm{Th}(\mathfrak{A})$, is the set of all sentences in the language of $\mathfrak{A}$ that are true in $\mathfrak{A}$. For each $\mathfrak{A}$, $\mathrm{Th}(\mathfrak{A})$ is a *complete theory*, i.e. for each sentence φ of the language of $\mathfrak{A}$, either φ is in $\mathrm{Th}(\mathfrak{A})$, or $\neg\varphi$ is in $\mathrm{Th}(\mathfrak{A})$. Clearly, $\mathfrak{A}$ is a model of $\mathrm{Th}(\mathfrak{A})$.

Can models be built for any theory? For some theories that cannot be done. If a set of sentences that contains a sentence and its negation it cannot have a model. Such a theory is *inconsistent*. It also cannot be done for theories that do not contain direct contradictions, but from which a contradiction can be derived by logical inference. Remarkably, it turns our that nonderivability of a contradiction is the only condition required for existence of a model of a theory. This was proved by Kurt Gödel in 1929. The result is known as Gödel's completeness theorem.

Theorem 11.1 *Let T be a first-order theory from which a contradiction cannot be formally derived. Then, there is a structure $\mathfrak{A}$ in which all sentences of T are true.*

Gödel proved his theorem for a particular deductive proof system, but it is valid for all systems satisfying certain natural conditions. It also has the following powerful corollary known as the compactness theorem.[3]

[1] For a sentence φ to logically follow from a set of sentences, means that there is a formal proof of φ in which sentences form the set are used as premises. Formal rules of proof have to be specified, and this can be done in several ways.

[2] In technical terms, the theory of a structure is axiomatizable if it has either finite, or a recursive (computable) set of axioms.

[3] Gödel proved the compactness theorem for countable languages. The theorem was extended to uncountable languages by Anatoly Maltsev in 1936.

Theorem 11.2 *If every finite set of sentences of a theory T has a model, then T has a model.*

While Theorem 11.1 is in fact many theorems, one theorem for each formal proof system, Theorem 11.2 is just one statement about theories and their models. Both theorems declare that for any set of first-order properties, as long as they don't contradict one another, there is a structure that has all those properties. What you can think of without contradictions, exists. But where and how do those structure exist? Mathematical theorems do not have the power of bringing objects to life. They are all about the realm of mathematical objects. The completeness and the compactness theorems are logical consequences of the axioms of set theory. They state that under certain assumptions certain sets and relations exist, and they have the required properties. In fact, the axioms of ZF alone are not strong enough to prove both theorems. In full generality, for arbitrary languages that may include uncountably many symbols, their proofs require the famous axiom of choice that declares that for every nonempty collection of disjoint sets there is a set that has exactly one element in common with each set in the collection. This seemingly innocuous axiom cannot be derived from the axioms of ZF, and it is necessary for proofs of many standard results in modern mathematics; hence, it is routinely included among the axioms and the resulting theory is abbreviated by ZFC.

Set theory is a commonly accepted formal framework for modern mathematics.[4] If it follows from the axioms of set theory that a certain set exists, then the set considered a bona fide mathematical object. For a working mathematician, it is as good as a circle drawn on paper. But one has to be cautious, since the set-theoretic approach opens up a whole world of fantastic objects that one can study, and that one becomes familiar in the process. Now and then it makes sense to stop and ask, Aren't we overdoing it? Perhaps the axioms we adopted are too strong. They allow to create a whole universe of objects with intriguing properties, but do we really need all of them? The experience of the last 100 years or so seems to indicate that the answer is yes, but still one should be aware that there may be a problem here.

What right do we have to claim that we gain real knowledge by investigating formal consequences of formal axioms? How do we know that the axioms are correct? How do we even know that they do not contradict each other? Early in the twentieth century several influential mathematicians and philosophers did raise such objections, to which David Hilbert gave his famous response: "No one will drive us from the paradise which Cantor created for us." [11] No one has driven us from the paradise. The new infinitistic methods proved to be extremely effective in all areas of mathematics. High levels of set-theoretic abstraction are used to prove results about much more concrete mathematical domains. A good example is Andrew Wiles' proof of Fermat's Last Theorem.[5]

---

[4]It is not the only formalism that is used in practice. A powerful alternative is category theory that is preferred in certain areas of algebra and geometry.

[5]For a very readable and comprehensive account see Simon Singh's book [31].

As for consistency of the axioms, after more that a century of intense and diverse investigations, no contradiction has ever been found, giving us a lot confidence, nevertheless the foundational debate is not over. One interesting aspect is the formal status of consistency of theories such as ZF or ZFC. If they are indeed consistent, then Hilbert believed that one should be able to provide a rigorous proof of it. Gödel's second incompleteness theorem, proved in 1931, implies that this is not possible. If a formal theory is strong enough, it cannot prove its own consistency, unless it is inconsistent. The qualification "strong enough" covers all formal theories that one might consider as a foundation base for all of mathematics. It seems that philosophical doubts will never be resolved by formal methods, and that the doubts may linger forever.

Just to give a flavor of the intensity of the foundational debate in the first half of the twentieth century, let me finish this section with a quote about Cantor's paradise from a famous philosopher:

> I would say. 'I wouldn't dream of trying to drive anyone out of this paradise.' I would try to do something quite different: I would try to show you that this is not a paradise—so that you'll leave of your own accord. I would say, 'You are welcome to this; just look about you.' [38]

11.2 New Structures from Old

The compactness theorem is one of the most powerful and frequently used results in modern mathematical logic. In this section, we will examine some of its applications. We will show that every structure with an infinite domain can be extended to a very similar structure with a larger domain. The notion of similarity that we use here will be defined in terms of first-order logic.

We can always expand a given structure to a larger one by adding new elements to its domain and by extending the relations of the structure to the new elements. That is easy, but we also want construct extensions in a way that preserves the character of the structure. If our structure is a graph, we want the extension to be a graph as well. This is not hard to do. However, if it is a field, and we want its extension to be a field as well it is a bit harder to do. It turns our that using the compactness theorem we can make all such extensions and much more without much effort. The theorem gives us a recipe for constructing *elementary* extensions that are defined below. The definition refers to types. See Definition 9.3.

Definition 11.1 Let $\mathfrak{A}$ and $\mathfrak{B}$ be structures for the same first-order language, with domains A and B respectively. If A is a subset of B, then we say that $\mathfrak{B}$ is an elementary extension of $\mathfrak{A}$ if the type of every tuple of elements of A is the same in $\mathfrak{A}$ and in $\mathfrak{B}$.

It follows directly from the definition that if $\mathfrak{B}$ is an elementary extension of $\mathfrak{A}$, then the theory of $\mathfrak{A}$ must be the same as the theory of $\mathfrak{B}$, and that is because the type of every tuple of elements of a structure along with formulas with free variables also contains all sentences that are true in the structure. This remark will provide us with many examples of extensions that are not elementary. In fact, in the process of extending the number structures we have already encountered them. We started with the domain of natural numbers $\mathbb{N}$ and extended it to the integers $\mathbb{Z}$, then to the rational numbers $\mathbb{Q}$, and finally to the real numbers $\mathbb{R}$. In the first-order language of addition and multiplication, none of those extensions are elementary. In fact, the theories of those structures are all different. This is precisely why we introduced those structures in the first place. In each larger structure we can solve some equations that have no solutions the smaller ones. Here are examples of first-order sentences that differentiate between them.

- $\forall x \exists y[x+y = 0]$ is false in $(\mathbb{N}, +, \cdot)$ and true in $(\mathbb{Z}, +, \cdot)$, $(\mathbb{Q}, +, \cdot)$, and $(\mathbb{R}, +, \cdot)$.
- $\forall x[(x = 0) \vee \exists y(x \cdot y = 1)]$ is false in $(\mathbb{Z}, +, \cdot)$ and true in $(\mathbb{Q}, +, \cdot)$ and $(\mathbb{R}, +, \cdot)$.
- $\exists x[x \cdot x = 2]$ is false in the $(\mathbb{Q}, +, \cdot)$ and true in $(\mathbb{R}, +, \cdot)$.[6]

Let us now see an example of an extension that is not elementary, but in which the structure and its extension share the same theory. Let $\mathbb{N}^+$ be the set of natural numbers without 0. Let $\mathfrak{A}$ be the ordered structure $(\mathbb{N}^+, <)$, and let $\mathfrak{B}$ be $(\mathbb{N}, <)$. $\mathfrak{B}$ is an extension of $\mathfrak{A}$, and it is an isomorphic copy of $\mathfrak{A}$. Both ordered sets look exactly the same. The function $f : \mathbb{N}^+ \longrightarrow \mathbb{N}$, defined by $f(n) = n - 1$, is an order preserving one-to-one correspondence between the two domains. Because the structures are isomorphic, their theories are the same, but the extension is not elementary. For example, 1 is the least element in $\mathfrak{A}$, and this property is first-order expressible, so it belongs to the type of 1. In $\mathfrak{B}$, 1 is no longer the least element, so the type of 1 has changed.

The following theorem is a consequence of the compactness theorem and implies that elementary extensions exist in abundance.

Theorem 11.3 *Every structure with an infinite domain has a proper*[7] *elementary extension.*

A proof of Theorem 11.3 is given in Appendix A.4. In the rest of this section we will discuss some powerful consequences.

Let us observe that the assumption that the domain of the structure in Theorem 11.3 is infinite is essential. If the domain of $\mathfrak{A}$ has a finite number of elements, for example 100, this fact can be expressed by a single first-order sentence. If $\mathfrak{B}$ is an elementary extension of $\mathfrak{A}$, then the same sentence must be true in $\mathfrak{B}$, hence the domain of $\mathfrak{B}$ must have 100 elements as well. Moreover, if $\mathfrak{A}$ and $\mathfrak{B}$ are structures with finite domains and $\mathrm{Th}(\mathfrak{A}) = \mathrm{Th}(\mathfrak{B})$, not only their domains of the structures must be of the same size, they must be isomorphic. This is an interesting and not entirely obvious fact. The proof of it is given in Appendix A.3.

[6]The parameters 0, 1, and 2 can be eliminated from these formal statements, since they are definable in each structure.

[7]Each structure is considered an extension of itself and it is an elementary extension. A proper extension is an extension that adds new elements to the domain of the extended structure.

In Theorem 11.3 we talk about structures with infinite domains rather then infinite structures. The reason is that a structure on a finite domain can be infinite, because it can have infinitely many relations. We have seen examples of such structures in Chap. 7.

Every structure with an infinite domain has a proper elementary extension, and that extension has an infinite domain, so it has a proper elementary extension, and so on. The process of creating elementary extensions can go on forever. And "forever" here means forever in the set-theoretic sense. We do not do, or create anything really. We use a metaphorical language to describe what set theory allows us to define formally, and to bring set-theoretic objects into being in this sense. Often, the number of steps in set-theoretic constructions is measured not by natural numbers, or rather not only by natural numbers. After the steps one, two, three, ..., comes the first limit step ω, and then ω plus one, ω plus two, and so on. The counting goes on into the transfinite, and the numbers that are used to count the steps, are *ordinal numbers*. The domain of each structure has its set-theoretic size that is its cardinality.

Theorem 11.3 has a sharper version that says that every structure $\mathfrak{A}$ with an infinite domain has a proper elementary extension with a domain of the same cardinality as the domain of $\mathfrak{A}$. In a proper extension, the domain gets enlarged, but set-theoretic size of the domain does not have to increase. However, iterating extensions sufficiently long, we can obtain elementary extensions of a given structure that are of arbitrarily large cardinality. For every structure there is a much larger structure that looks very much like it. Let us see how it works in an example.

The ordered set of the rational numbers $(\mathbb{Q}, <)$ is a dense linear ordering without endpoints, and this fact is expressible by a single first-order sentence in the language with the relation symbol $<$. Let $(D, <)$ be a proper elementary extension of $(\mathbb{Q}, <)$. Since the theory of $(D, <)$ is the same as the theory of $(\mathbb{Q}, <)$, the sentence expressing that the ordering is dense and has no endpoints is also true in $(D, <)$; hence $(D, <)$ is a dense linear ordering without endpoints. Since $\mathbb{Q}$ is countable we can assume that D is countable as well. Any two countable densely ordered sets without endpoints are isomorphic. This means that $(D, <)$ is an isomorphic copy of $(\mathbb{Q}, <)$. It looks exactly the same. So in this case, even though Theorem 11.3 allows us to construct an extension with a larger domain,[8] we are not really getting a genuinely new structure. However, if we keep making the extensions long enough, if we "go through" uncountably many steps, we will have added uncountably many new elements to the domain, and this will give us densely ordered set of uncountable size. By itself, this is not a great achievement. $(\mathbb{R}, <)$ is also an uncountable dense linear ordering without endpoints, but what is interesting here is not only that the process can be continued forever, creating dense linear orderings of arbitrarily large sizes. Constructions of long chains of elementary extensions can be carried out in many substantially different ways, resulting in many similar but non-isomorphic structures. For example, we can start with $(\mathbb{Q}, <)$ and proceed in such a way that the

[8]Larger in the sense of containment of sets, not their cardinalities.

new elements are always larger than all old elements. If this is done over continuum many steps,[9] then the resulting structure is a densely ordered set of cardinality continuum that is not isomorphic to $(\mathbb{R}, <)$.

11.2.1 Twin Primes

We will finish this chapter with an application of the compactness theorem to a number-theoretic problem.

A pair of *twin primes* is a pair of prime numbers that differ by 2. Here are the first six twin prime pairs: (3,5), (11,13), (17,19), (29,31), (41, 43), (59,61). As we go up the ladder of natural numbers we sporadically encounter twin primes. Since there are infinitely many prime numbers, one can ask whether there are also infinitely many twin primes. The Twin Primes Conjecture says that there are, but despite serious efforts, it has not been confirmed.[10]

Here is how one could try to approach the conjecture using model theory. Let $P(x)$ be a first-order formula defining the set of prime numbers in $(\mathbb{N}, +, \cdot)$. The sentence

$$\forall x \exists y [x < y \wedge P(y) \wedge P(y+2)]$$

formally expresses the Twin Primes Conjecture. Let us call this sentence TPC.[11]

The compactness theorem not only tells us that $(\mathbb{N}, +, \cdot)$ has a proper elementary extension, it also can be used to show that there is a great variety of such extensions. Since all those extensions are elementary, each has an unbounded set of elements with the property expressed by the formula $P(x)$. In each extension all old prime numbers in $\mathbb{N}$ still have property $P(x)$, but there are also some new elements c for which $P(c)$ holds. Let us call them *nonstandard primes*. Suppose now that in some extension $(\mathbb{N}^*, +, \cdot)$, we can find a nonstandard prime c such that $c + 2$ is also a prime. Then for every (standard) natural number n the following sentence is true in $(\mathbb{N}^*, +, \cdot)$

$$\exists y [n < y \wedge P(y) \wedge P(y+2)].$$

This sentence is true in $(\mathbb{N}^*, +, \cdot)$, because c is such a y. But this sentence is not referring to any new element in the extension, hence—because the extension is

[9] Continuum is the cardinal number of the set of real numbers $\mathbb{R}$.

[10] According to Wikipedia, the current largest twin prime pair known is $2996863034895 \cdot 2^{1290000} \pm 1$, with 388,342 decimal digits. It was discovered in September 2016.

[11] Some small alterations are needed to make TPC comply with the first-order formalism. To express TPC as a first-order sentence in the language of $+$ and $\cdot$, one has to replace $<$ by its definition in $(\mathbb{N}, +, \cdot)$, the expression $P(y + 2)$ can be written as $\forall z[(z = y + 2) \implies P(z)]$, and the reference to 2 can be eliminated with the help of its definition in $(\mathbb{N}, +, \cdot)$.

elementary—it must be also true in $(\mathbb{N}, +, \cdot)$. It follows that for any natural number n there are twin primes in $\mathbb{N}$ that are bigger than n, and that means that there are infinitely many (standard) twin primes.

By considering elementary extensions, the task of proving that there are infinitely many twin primes, got reduced to the task of finding one nonstandard model with just one pair of nonstandard twin primes. One could hope to solve an outstanding problem in number theory by constructing a particular structure. Unfortunately, it is not as easy as it seems. Recent work in number theory suggests that while the conjecture is likely to be true, any proof of it will be very difficult. If the conjecture is in fact true, then in any elementary extension of $(\mathbb{N}, +, \cdot)$ there are infinitely many nonstandard twin primes, but most likely the only way to prove it is by actually proving the conjecture first.

Exercises

Exercise 11.1 *Write the axiom of choice as a first-order statement using only the membership relation symbol* $\in$.

Exercise 11.2 *Suppose that* $\mathfrak{B}$ *is an elementary extension of* $\mathfrak{A}$. *Show that every relation of* $\mathfrak{B}$ *restricted to the domain of* $\mathfrak{A}$ *is one of the relations of* $\mathfrak{A}$. *Hint: This follows directly from Definition* 11.1.

Exercise 11.3 **Prove Cantor's theorem: if* $(D, <)$ *and* $(E, <)$ *are countable dense linearly ordered sets without endpoints, then they are isomorphic. Hint: Assume that* $D = \{d_1, d_2, \dots\}$ *and* $E = \{e_1, e_2, \dots\}$. *Then, in a step-by-step fashion, construct a one-to-one correspondence between D and E preserving the ordering of the sets.*

Exercise 11.4 *Show that there are infinitely many types of pairs of elements in* $(\mathbb{Z}, <)$. *Hint: For integer numbers a and b, with* $a < b$, *consider the distance between a and b, i.e. the number of integers x such that* $a < x + 1 < b$. *Show that for each natural number n, the relation "the distance between a and b is n" is definable in* $(\mathbb{Z}, <)$.

Exercise 11.5 *Show that if a, b, c, and d are rational numbers, and* $a < b$ *and* $c < d$, *then the types of* (a, b) *and* (c, d) *in* $(\mathbb{Q}, <)$ *are equal. Hint: Find a symmetry* $f : \mathbb{Q} \longrightarrow \mathbb{Q}$ *such that* $f(a) = c$ *and* $f(b) = d$.

Exercise 11.6 ** Show that there are nonisomorphic dense linear orderings of power continuum. Hint: A construction is outlined in Sect.* 11.2.

Chapter 12
Elementary Extensions and Symmetries

Abstract In this chapter we will see how one can learn something about a structure by using symmetries of its elementary extensions. We will examine the specific example of the ordering of the natural numbers, and we will prove that the structure $(\mathbb{N}, <)$ is minimal. After so many pages, the reader will probably find it hard to believe that this example was my original motivation to write this book. Initially, it seemed that not much technical preparation was needed.

Keywords Minimality · Symmetries of the ordered set of integers · Ehrenfeucht-Mostowski theorem · Ramsey type theorems · Homogeneous sets

12.1 Minimality of $(\mathbb{N}, <)$

The structure we will talk about is the ordered set of natural numbers $(\mathbb{N}, <)$. It is easy to visualize. No advanced mathematics is involved. Still, what we will do is not trivial, and we will use much of the power of first-order logic.

To say that we will learn something about the structure of $(\mathbb{N}, <)$ is an exaggeration. $(\mathbb{N}, <)$ is such a simple structure that there is nothing really that we need to learn about it. All one may want to know is already shown in the image of a sequence of points starting at 0, going up or, as the number line is usually drawn, going to the right, and then disappearing into infinity.

$$\bullet \quad \bullet \quad \bullet \quad \bullet \quad \bullet \cdots$$

We will not learn anything new about this structure but we will learn something nontrivial about its definable sets; hence we will learn something about logic. Our goal is to characterize completely all sets of natural numbers that can be defined in $(\mathbb{N}, <)$. The structure looks simple, but we have a language in which we can express complicated properties, so it is hard to say in advance what can and what cannot be defined.

R. Kossak, *Mathematical Logic*, Springer Graduate Texts in Philosophy 3,
https://doi.org/10.1007/978-3-319-97298-5_12

What may be unclear about the that simple image of $(\mathbb{N}, <)$ is the role of the three small dots on the right. They indicate that the points go to infinity and they do it in an orderly fashion. But how do they do it? Where does the number line go? It is good to have a picture in mind, especially since the reader will be soon asked to think of new points that will be appended at the right, beyond all those infinitely many points. Another, more concrete representation may help.

Let E be the set of points on the number line corresponding to the numbers

$$1 - \frac{1}{2},\ 1 - \frac{1}{3},\ 1 - \frac{1}{4}, 1 - \frac{1}{5}, \ldots$$

The sequence begins with $1 - \frac{1}{2} = \frac{1}{2}$, and then proceeds up on the number line, getting closer and closer to 1, but never reaching it. As ordered sets, $(\mathbb{N}, <)$ and $(E, <)$ are isomorphic. They are two representations of same structure, but while $(\mathbb{N}, <)$ can be viewed as an unbounded subset of the real line, E is bounded, so we will have no problem visualizing how new elements can be added to it. Here is the picture of E with one new point, the number 1, added to it on top:

$$1 - \frac{1}{2},\ 1 - \frac{1}{3},\ 1 - \frac{1}{4}, 1 - \frac{1}{5}, \ldots, 1$$

If we now forget that the elements of listed above are numbers, and we only record how they are ordered, we get a simple picture:

$$\bullet \quad \bullet \quad \bullet \quad \bullet \cdots \bullet$$

This is really all we need. The representation using fractions helps to see how an infinite sequence of elements can form an increasing chain, but still be bounded from above.

Now let us take a look at the ordered set $(\mathbb{Z}, <)$. It looks like $(\mathbb{N}, <)$, except that it is unbounded at both ends.

$$\cdots \bullet \quad \bullet \quad \bullet \quad \bullet \quad \bullet \cdots$$

For a more concrete representation, we can use the ordered set F consisting of the numbers

$$\ldots, \frac{1}{4}, \frac{1}{3}, \frac{1}{2}, 1,\ 2 - \frac{1}{2},\ 2 - \frac{1}{3},\ 2 - \frac{1}{4}, \ldots$$

In this representation, 1 plays the role of the midpoint from which other points move to the left towards 0, and to the right towards 2. All numbers in F are between 0 and 2, approaching those two numbers arbitrarily closely, but never reaching them.

The structure $(E, <)$ is an isomorphic copy of $(\mathbb{N}, <)$, and $(F, <)$ is an isomorphic copy of $(\mathbb{Z}, <)$. Now we can form a new structure whose domain will be the union of E and F, and the ordering relation is the usual ordering on the number line. Here is a picture

$$\bullet \quad \bullet \quad \bullet \quad \bullet \quad \bullet \cdots \qquad \cdots \bullet \quad \bullet \quad \bullet \quad \bullet \quad \bullet \cdots$$

This new ordered structure is isomorphic to $(\mathbb{N}, <)$ with an isomorphic copy of $(\mathbb{Z}, <)$ added on top. The only reason to consider E and F instead of $\mathbb{N}$ and $\mathbb{Z}$ is to make this extended structure easier to visualize.

$(\mathbb{N}, <)$ has a least element, and $(\mathbb{Z}, <)$ does not. This is the only essential difference, but it has further consequences. As we have seen, in $(\mathbb{N}, <)$ every element is definable.

Recall that a symmetry of a structure is a permutation of its domain that preserves all relations of the structure, and as a consequence it also preserves all first-order properties of tuples of elements of the domain. In particular if f is a symmetry of a structure then, for every element a of the domain, the type of a is the same as the type of its image $f(a)$. Since in $(\mathbb{N}, <)$ every element is definable, each element has its own unique type, and it follows that $(\mathbb{N}, <)$ has no symmetries at all. In a stark contrast, all elements of $\mathbb{Z}$ share the same type. Let us see why.

We will repeat an argument we have already used in Sect. 9.4. Let a and b be integers, and let us assume that $a < b$. Let $f: \mathbb{Z} \longrightarrow \mathbb{Z}$ be the permutation that shifts all elements up by $b - a$, i.e. $f(x) = x + (b - a)$. Then, since f preserves the ordering, it is a symmetry of $(\mathbb{Z}, <)$, and because $f(a) = b$, the type of a is the same as the type of b. While in $(\mathbb{N}, <)$ every element has its own unique type, in $(\mathbb{Z}, <)$, any element looks exactly as any other element.

We are now ready for a more involved argument that will show that $(\mathbb{N}, <)$ is minimal.

Suppose that $(\mathbb{N}, <)$ is not minimal. Then, there is a formula $\varphi(x)$ of the first-order language with one binary relation symbol $<$ defining a subset of $\mathbb{N}$ that is neither finite not cofinite. That means that the following holds in $(\mathbb{N}, <)$

$$\forall x \exists y \exists z[(x < y) \wedge (y < z) \wedge \varphi(y) \wedge \neg\varphi(z))]. \qquad (*)$$

We have seen that the relation "y is a successor of x" is definable in $(\mathbb{N}, <)$. Let $S(x, y)$ be a formula defining it, i.e. for all natural numbers m and n, $S(m, n)$ holds in $(\mathbb{N}, <)$ if and only if $n = m + 1$. Then, there are infinitely many n such $\varphi(n)$ holds and $\varphi(n + 1)$ does not. In symbols:

$$\forall x \exists y \exists z[(x < y) \wedge S(y, z) \wedge \varphi(y) \wedge \neg\varphi(z))]. \qquad (**)$$

By the compactness theorem, $(\mathbb{N}, <)$ has a proper elementary extension $(\mathbb{N}^*, <)$. Since the extension is proper, it has a new element c, and this element must be larger than all natural numbers in $\mathbb{N}$. Why? Consider a natural number, for example 50. Could c be smaller than 50? Let us see. The following sentence holds in $(\mathbb{N}, <)$

$$\forall x[x < 50 \Longrightarrow (x = 0 \vee x = 1 \vee x = 2 \vee \cdots \vee x = 49)].$$

The above sentence expresses the fact that every natural number smaller than 50 must be one of the numbers from 0 to 49. Since the extension is elementary, the same sentence holds in $(\mathbb{N}^*, <)$ and since c is a new element, it cannot be smaller than 50.

Since every element of $(\mathbb{N}, <)$ has a successor, the same is true in $(\mathbb{N}^*, <)$. So c has a successor $c+1$, and $c+1$ has a successor $c+2$, and so on. Also, in $\mathbb{N}$, every element, except for 0, has a predecessor. This is expressed formally by

$$\forall x[\neg(x = 0) \Longrightarrow \exists y\, S(y, x)].$$

The same holds in $(\mathbb{N}^*, <)$, and it implies that c has a predecessor, let us call it $c-1$. This $c-1$ is a also a new element. If $c-1$ were an old natural number, for example 23, then c would be 24, and we know already that it can't be. So $c-1$ is new, hence it has a predecessor $c-2$, which is also new, and so on.

The argument above shows that $\mathbb{N}^*$ has many other new elements, and that every new element, such as the c above, must be a part of a predecessor/successor chain of elements that looks exactly like $(\mathbb{Z}, <)$.

Now comes a crucial point in the proof. Because $(**)$ holds in $(\mathbb{N}^*, <)$, there infinitely many new elements c such that

$$\varphi(c) \wedge \neg\varphi(c+1) \qquad (***)$$

holds in $(\mathbb{N}^*, <)$.[1] Let us fix such a c.

Now we can define a symmetry of $(\mathbb{N}^*, <)$ as follows. For all elements a that are not in the predecessor/successor chain of the c above, we let $f(a) = a$. Those elements don't move. For every element a in the chain of c, let $f(a)$ be $a+1$. In particular, $f(c) = c+1$. Since for all a and b, $a < b$ if and only if $f(a) < f(b)$, f is a symmetry of $(\mathbb{N}^*, <)$. Because f is a symmetry, c and $f(c)$ must have the same type in $(\mathbb{N}^*, <)$, it follows that, $\varphi(c)$ holds in $(\mathbb{N}^*, <)$ if and only $\varphi(c+1)$ does, but we chose c so that only $\varphi(c)$ holds, so this is a contradiction. We have proved that $(\mathbb{N}, <)$ is minimal.

There are several aspects of the proof we just saw that are worth stressing. First of all, we carried out the whole argument not knowing much about $(\mathbb{N}^*, <)$. The compactness theorem just tells us that it exists, but it tells us little about what it looks like, except that is an elementary extension of $(\mathbb{N}, <)$. Once we know that the extension is elementary, the argument rests on our ability to express relevant properties of $(\mathbb{N}, <)$ in a first-order way, to be able to transfer them and use in the extension $(\mathbb{N}^*, <)$.

The second important aspect is the use of symmetry. We cannot use symmetries directly to argue that certain relations are not definable in $(\mathbb{N}, <)$ because it is a rigid structure, it has no nontrivial symmetries at all. To take advantage of symmetries we

---

[1]We are using the symbol + here. It is not in the language of $(\mathbb{N}, <)$, but it is allowed as an abbreviation, since the expression $\varphi(c+1)$ can be written as $\forall z S(c, z) \Longrightarrow \varphi(z)$.

had to move to a larger structure that has them. This may seem as a rather ad hoc trick, but in fact it is a standard method of model theory. It is widely applicable partly due to the fact that every structure with infinite domain has a proper elementary extension that admits nontrivial symmetries. This was proved in 1956 by Andrzej Ehrenfeucht and Andrzej Mostowski. The theorem of Ehrenfeucht and Mostowski is too advanced to be included here with a proof, but in the next section some relevant details are mentioned.

12.2 Building Symmetries

Here is a somewhat curious fact of life. At any party that is attended by six or more people, there will always be either at least three mutual acquaintance or at least three mutual strangers. To prove it, let us consider six people A, B, C, D, E, and F. We assume nothing specific about who knows whom, in particular they could all know each other, or all be mutual strangers. In both cases, surely our statement is true. In the first case we do have six mutual acquaintances, in the second six mutual strangers. We could try to check if the same is true for all other cases, but this would be a tedious task. Six people can be paired in 15 ways,[2] and each pair can potentially be a pair of acquaintances or strangers. This gives us $2^{15} = 32{,}768$ possible relations to verify. Instead, we will use a clever argument.

Assume nothing specific about who knows whom, and consider the person A. The five remaining people are split into two sets, in the first are the people that A knows, and in the second everybody else. Regardless of whom A knows, one of those sets must include at least three people. If both had less than three members, then the total number of members in both would be less than 5, and that can't be. Simple.

Suppose now that three people, say B, C and D, are in the first set. The argument is similar if three people are in the second set. If any of the B, C, or D know each other, that, together with A, creates a triangle of three people who know each other. If not, then B, C, and D form a triangle of strangers, and this proves our statement.

Number six is the smallest number with the property just described. In a group of five people it can happen that neither three people know each other nor there are three mutual strangers.

What we proved about acquaintances and strangers can be formulated in terms or relations as follows.

Theorem 12.1 *Let $\mathfrak{A}$ be a structure whose domain has at least six elements, with a binary relation E such that*

$$\forall x \forall y\, [(E(x, y) \Longrightarrow E(y, x)) \land \forall x\, \neg E(x, x)].$$

[2]This follows from a general fact that the number of pairs in an n-element set is $\frac{n(n-1)}{2}$. For small values of n such as 6, one can verify it by listing all possible pairs.

Then there is a subset X of the domain such X has at least three elements, and either for all distinct a, and b in X, $E(a, b)$ holds, or for all distinct a, and b in X, $\neg E(a, b)$ holds. Such a set X is called homogenous with respect to E.

Theorem 12.1 is an example of a Ramsey type theorem.[3] All Ramsey type theorems have a common structure, they say that for given n and k, and a given property of k-tuples, if the domain a structure is large enough, than it has a subset X of size at least n, that is homogeneous with respect to the property, i.e. either all k-tuples of elements of X have the property, or all of them don't. In other words, all k-tuples in X look the same with respect to the property. Notice that this does not mean that all k-tuples of X all have the same type. The structure may have other relations, and even if it does not, the elements of X may interact with elements of the structure outside X in different ways, hence they types can be different. Nevertheless, Ramsey type theorems tell us that for any given property, if the domain of a structure is large enough, then it must have many elements that "look alike" with respect to the given property.

There is also a powerful infinitary Ramsey's theorem. Here is a variant of it. Suppose $\mathfrak{N} = (\mathbb{N}, <, \dots)$ is a structure, where $<$ is the usual ordering of the natural numbers, and the dots indicate that the structure may or may not have other relations. Let $\varphi(x_1, x_2, \dots, x_k)$ be a formula of the language of $\mathfrak{N}$. For every infinite set X that is definable in $\mathfrak{N}$, there is an infinite definable Y contained in X that is homogenous for the property defined by $\varphi(x_1, x_2, \dots, x_n)$ in the sense that either for all increasing sequences $a_1 < a_2 < \cdots < a_k$ of elements of Y, $\varphi(a_1, a_2, \dots, a_k)$ holds, or for all such sequences $\neg\varphi(a_1, a_2, \dots, a_k)$ holds. This means that while $\mathfrak{N}$ may be rigid, for any finite number of first-order properties $\mathfrak{N}$ there is an infinite definable set whose elements all "look alike" with respect to those properties. This theorem and the compactness theorem are the main ingredients of Ehrenfeucht and Mostowski's theorem on existence of elementary extensions with symmetries. A proof of the infinitary version of Ramsey's theorem for pairs ($k = 2$) is given in Appendix A.5.

Exercises

Exercise 12.1 *Show that the that there is no formula $\varphi(x, y, z)$ of the language with one binary relation symbol $<$ such that for all natural numbers k, l, m, $\varphi(k, l, m)$ holds in* $(\mathbb{N}, <)$ *if and only if $k + l = m$. Hint: There are many ways in which this can be shown. For a short argument, think of the set of even numbers, and use minimality of* $(\mathbb{N}, <)$.

Exercise 12.2 *Use the previous exercise to show that the addition of natural numbers is not definable in the structure* $(\mathbb{N}, S)$, *where S is the successor relation, i.e. for all natural numbers m and n, $S(m, n)$ holds if and only if $m + 1 = n$.*

[3] After British philosopher, mathematician, and economist Frank Plumpton Ramsey (1903–1930).

Chapter 13
Tame vs. Wild

Abstract In this chapter we will compare two classical structures: the field of complex numbers $(\mathbb{C}, +, \cdot)$ and the standard model of arithmetic $(\mathbb{N}, +, \cdot)$. The former is vast and mysterious, the latter deceptively simple. As it turns out, as far as the model-theoretic properties of both structures are concerned, the roles are reversed, the former is very tame while the latter quite wild, and those terms have well-understood meanings. In recent years, tameness has become a popular word in model theory. Tameness is not defined formally, but a structure is considered tame if the geometry of its definable sets is well-described and understood. Tameness has different levels. The most tame structures are the minimal ones. All parametrically definable unary relations in a minimal structure are either finite or cofinite. The examples of minimal structures that we have seen so far are the structures with no relations on them—the trivial structures—and $(\mathbb{N}, <)$. It is somewhat surprising that the ultimate number structure—the complex numbers, is also minimal. It is a fascinating example.

Keywords Complex numbers · Cantor's pairing function · Arithmetization of language · Tarski's undefinability of truth theorem · Gödel's second incompleteness theorem

13.1 Complex Numbers

In order to extend the field of the rational numbers to include numbers such as $\sqrt{2}$, π, and e, we employed the concept of Dedekind cut, and accepted all set-theoretic complexities it involved. Now we will enlarge the domain of numbers one more time, but this move will be easier. The set $\mathbb{C}$ of complex numbers is simply $\mathbb{R}^2$, the set of all ordered pairs of real numbers.[1] To make complex numbers numbers, we need to define how they are added and multiplied.

[1] Complex numbers are often defined as expressions of the form $a + bi$, where a and b are real number, and i is such that $i^2 = -1$. You will see below that our definition is equivalent.

R. Kossak, *Mathematical Logic*, Springer Graduate Texts in Philosophy 3,
https://doi.org/10.1007/978-3-319-97298-5_13

Definition 13.1 For the complex numbers (a, b) and (c, d)

$$(a, b) + (c, d) = (a + c, b + d)$$

and

$$(a, b) \cdot (c, d) = (ac - bd, ad + bc).$$

Notice that addition and multiplication of complex numbers are definable in $(\mathbb{R}, +, \cdot)$. It follows that the logical complexity of the structure $(\mathbb{C}, +, \cdot)$ is no greater than that of $(\mathbb{R}, +, \cdot)$. The geometry of the complex numbers is a part of the geometry of the real numbers.

Addition of complex numbers is straightforward, but why is multiplication so strange? The answer is: Because it works. With this definition $(\mathbb{C}, +, \cdot)$ becomes a field. And what a field it is!

First, let us see why $\mathbb{C}$ can be considered an extension of $\mathbb{R}$. Indeed, each real number r can be identified with the pair $(r, 0)$, and it is easy to check that addition and multiplication of such pairs agrees with addition and multiplication of the corresponding real numbers. Thus $\mathbb{C}$ contains a copy of $\mathbb{R}$, and in this sense we consider it an extension. Also, for this reason, for a complex number (a, b), a is called its *real part*, and b its *imaginary part*.

Let i be the complex number $(0, 1)$. It is called the *imaginary unit*. Let us compute $i^2 = (0, 1) \cdot (0, 1)$. According to Definition 13.1,

$$(0, 1) \cdot (0, 1) = (0 \cdot 0 - 1 \cdot 1, 0 \cdot 1 + 1 \cdot 0) = (-1, 0).$$

Since we identified $(-1, 0)$ with -1, we can say that $i^2 = -1$. It follows that i is a solution of the equation $x^2 + 1 = 0$. This equation has another solution in $\mathbb{C}$. It is $-i = (0, -1)$, as can be easily checked. Thus the sentence $\exists x\ [x^2 + 1 = 0]$ is true in $(\mathbb{C}, +, \cdot)$, and false $(\mathbb{R}, +, \cdot)$. This shows that $(\mathbb{C}, +, \cdot)$ is not an elementary extension of $(\mathbb{R}, +, \cdot)$.

Not only $x^2 + 1 = 0$ has solutions in $\mathbb{C}$. All polynomial equations have. The field $(\mathbb{C}, +, \cdot)$ is *algebraically closed*. Any polynomial equation with coefficients in $\mathbb{C}$ has a solution in $\mathbb{C}$. This nontrivial fact, called the Fundamental Theorem of Algebra, was first proved rigorously by Jean-Robert Argand in 1806.

More polynomial equations are solvable in $\mathbb{R}$ than in $\mathbb{Q}$, and this is why the field $(\mathbb{R}, +, \cdot)$ is tamer than $(\mathbb{Q}, +, \cdot)$. In $\mathbb{C}$ all polynomial equations have solutions, and the structure becomes as tame as possible. The field of complex numbers is minimal. This is a consequence of a theorem of Chevalley and Tarski, who (independently) proved that a projection of a set that is definable in $(\mathbb{C}, +, \cdot)$ by a polynomial equation is itself definable by a polynomial equation. This allows to show that the field of complex numbers admits elimination of quantifiers, and from this it follows that every parametrically definable subset of $\mathbb{C}$ is either finite or cofinite.

13.1.1 Real Numbers and Order-Minimality

More complex, but still tame, are the order-minimal structures. Those are the structures whose domains are linearly ordered by a definable binary relation, and all of whose parametrically definable subsets of the domain are finite unions of intervals. $(\mathbb{Q}, <)$, $(\mathbb{Q}, +, <)$, and $(\mathbb{R}, +, \cdot, <)$ are order-minimal, as are various other structures obtained by expanding $(\mathbb{R}, +, \cdot, <)$ by adding functions and relations to it. In 1991, Alex Wilkie solved an old problem posed by Tarski, by showing that $(\mathbb{R}, +, \cdot, <, \exp)$, where exp is the binary relation $y = 2^x$, is also order-minimal. Wilkie's result initiated a whole area of study of order-minimal expansions of the field of real numbers. There is a beautiful theorem due to Ya'acov Peterzil and Sergei Starchenko, which states that the structures listed above are in a sense the only order-minimal structures.[2]

There are other levels of tameness, and there is a well-developed theory of tame structures with applications in classical algebra and analysis. We cannot go into more details. Now it is time to see what is on the other side, where wild structures live.

13.2 On the Wild Side

The additive structure of the natural numbers $(\mathbb{N}, +)$ and the multiplicative structure $(\mathbb{N}, \cdot)$ are considered simple. Neither is minimal. The formula $\exists y(y+y = x)$ defines the set of even numbers in $(\mathbb{N}, +)$, and $\exists y(y \cdot y = x)$, the set of square numbers in $(\mathbb{N}, \cdot)$, and those sets are neither finite nor cofinite. Nevertheless, parametrically definable sets in $(\mathbb{N}, +)$ and $(\mathbb{N}, \cdot)$ are well-understood, hence tame. All hell breaks loose when the two operations are combined in the *standard model of arithmetic* $(\mathbb{N}, +, \cdot)$. We will denote $(\mathbb{N}, +, \cdot)$ by $\mathfrak{N}$.

Much of the elegant theory of minimal and order-minimal structures is based on considerations involving a notion of dimension for definable sets. No such notion is available for $\mathfrak{N}$.

In set theory, one can prove that for every infinite set X, there is a one-to-one correspondence between the set of pairs X^2 and X, i.e. there is a one-to-one and onto function $f : X^2 \longrightarrow X$. Such a function always exists, but it may not be easy to find an explicit definition for it. Georg Cantor noticed that for the set $\mathbb{N}$ there is such a function with a particularly simple definition. He defined $C : \mathbb{N}^2 \longrightarrow \mathbb{N}$ as follows

$$C(x, y) = \frac{1}{2}(x + y + 1) \cdot (x + y) + y.$$

[2]For the precise statement of the theorem and an interesting discussion see [28].

Relation $C(x, y) = z$ is definable in $\mathfrak{N}$.[3] It is a small technical fact, but it has important consequences. Speaking metaphorically, first-order logic of $\mathfrak{N}$ sees that $\mathbb{N}^2$ and $\mathbb{N}$ have the same size. This implies that there can be no good notion of dimension for sets which are definable over $\mathfrak{N}$. Every definable subset of $\mathbb{N}^n$, for any n, can be coded in a definable way by a subset of $\mathbb{N}$. The geometry of $\mathfrak{N}$ is flat. Nothing like that can happen in $(\mathbb{R}, +, \cdot)$ nor in $(\mathbb{C}, +, \cdot)$, no one-to-one correspondence between the domain and its Cartesian square is definable in these structures.

Nothing too wild is happening yet, but the fact that higher dimensions are compressed to one in $\mathfrak{N}$ is just a beginning. In 1931, Kurt Gödel proved that the theory of $\mathfrak{N}$ is not axiomatizable, which means that for any effectively presented set axioms for $\mathfrak{N}$, there is a first-order sentence that is true in $\mathfrak{N}$, but does not logically follow from the axioms. In the proof, Gödel exposed the great expressive power of the first-order language of arithmetic. It followed from his analysis that any set of natural numbers that can be generated by an effective process has a first-order definition in $\mathfrak{N}$. All computable sets of natural numbers are definable, and many noncomputable sets are definable as well.

An important ingredient of Gödel's proof is the "arithmetization of language." Each formal symbol of the first-order language of arithmetic is assigned a natural number that serves as its code. Then, using Cantor's coding of finite sequences each formula is coded by a single number known as its *Gödel number*. If T is an effectively presented set of axioms for $\mathfrak{N}$, then the set of Gödel numbers of the sentences in T is definable in $\mathfrak{N}$, and so is the set of their logical consequences $Cons_T$. Moreover, if T is strong enough, for example if it is the set of Peano's axioms, then $Cons_T$ is not computable. This gives us examples of definable noncomputable sets. But there is more. For every number n, the set of Gödel numbers of sentences with no more than n quantifiers that are true in $\mathfrak{N}$ is definable, but, as Alfred Tarski showed, the set of Gödel numbers of all sentences that are true in $\mathfrak{N}$ is not—this is the famous Tarski's undefinability of truth theorem. A closely related argument shows that for every $n > 0$ there are sets of natural numbers that have definitions involving n quantifiers, but cannot be defined by a formula with fewer than n quantifiers. This is already a pretty wild picture, but it perhaps takes a mathematically trained eye to see that. Here comes something even more radical.

It has been already mentioned that the question whether the formal theory ZFC is consistent is delicate. Most mathematicians believe that it is, but we also know that, due to Gödel's Second Incompleteness theorem, we cannot prove that it is consistent within the theory itself. Consequently, when we talk about consistency of ZFC we have to treat it as a plausible, but unproven conjecture. If ZFC is consistent, then by Gödel's completeness theorem, there is a set V and a binary relation E on it, which is a model of the ZFC axioms. The completeness theorem has its arithmetized version which says that if an effectively presented set of axioms is consistent, then it has a model whose domain is $\mathbb{N}$ and whose relations are definable in $\mathfrak{N}$. Since

[3]The defining formula is $z + z = (x + y + 1) \cdot (x + y) + y$.

the axioms of **ZFC** are effectively presented, if **ZFC** is consistent, then there is a binary relation E on $\mathbb{N}$, definable in $\mathfrak{N}$, such that $(\mathbb{N}, E)$ is a model of **ZFC**. This is unexpected. Often in this book I stressed the enormous power of set theory. In **ZFC** we can reconstruct most of mathematics. All classical mathematical objects can be modeled as sets constructed using the axioms. Now it turns out that all those objects can be found as natural numbers (!) in $(\mathbb{N}, E)$. Now, this is wild.

We all learned in school that multiplication is repeated addition: $5 \cdot 2 = 2+2+2+2+2$. The addition of natural numbers completely determines their multiplication. An interesting corollary of the results that we discussed above is that multiplication is not definable in $(\mathbb{N}, +)$. If it were, this would allow us to reconstruct the whole wilderness of $\mathfrak{N}$ inside $(\mathbb{N}, +)$ and, because the latter structure is somewhat tame, this cannot be done.

Since $\mathfrak{N}$ is wild, any structure in which one can define an isomorphic copy of $\mathfrak{N}$ is wild as well. For example, $\mathfrak{N}$ can be defined in $(\mathbb{Z}, +, \cdot)$. The definition is not complicated, but it uses a nontrivial number-theoretic fact. A theorem of Lagrange states that every natural number can be written as a sum of four squares.[4] Hence, an integer n is nonnegative if and only if there are numbers n_1, n_2, n_3, and n_4 such that

$$n = n_1^2 + n_2^2 + n_3^2 + n_4^2.$$

It follows that the formula

$$\exists x_1 \exists x_2 \exists x_3 \exists x_4 [x = x_1^2 + x_2^2 + x_3^2 + x_4^2]$$

defines the set of nonnegative integers $\mathbb{N}$ in $(\mathbb{Z}, +, \cdot)$.

The natural numbers can also be defined in $(\mathbb{Q}, +, \cdot)$. It is a theorem of Julia Robinson, and we discussed it in Chap. 10.

Exercises

Exercise 13.1 *Use Definition* 13.1 *to verify that the distributive property* $a \cdot (b + c) = a \cdot b + a \cdot c$ *holds for all complex numbers a, b, and c.*

Exercise 13.2 *Write first-order formulas defining addition and multiplication of complex numbers in* $(\mathbb{R}, +, \cdot)$. *Hint: Remember that complex numbers are defined as pairs of real numbers.*

Exercise 13.3 *Let* $C : \mathbb{N}^2 \longrightarrow \mathbb{N}$ *be Cantor's pairing function. Compute* $C(0, 0)$, $C(1, 0)$, $C(0, 1)$ *and* $C(1, 1)$.

[4]For example $5 = 2^2 + 1^2 + 0^2 + 0^2$, $12 = 3^3 + 1^1 + 1^2 + 1^2$, and $98 = 9^2 + 3^2 + 2^2 + 2^2$.

Exercise 13.4 *Cantor's pairing function C can be used to define a one-to-one correspondence between $\mathbb{N}^n$ and $\mathbb{N}$, for any $n > 2$. For example, we can define $D : \mathbb{N}^3 \longrightarrow \mathbb{N}$ by $D(x, y, z) = C(x, C(y, z))$. Show that the relation $D(x, y, z) = w$ is definable in* $(\mathbb{N}, +, \cdot)$.

Chapter 14
First-Order Properties

Abstract A first-order property of a structure is a property that can be expressed in first-order logic. Some important properties are first-order but many are not. We will see why finiteness, minimality, order-minimality, and being well ordered are not first-order, and how some such properties can be expressed in higher-order logics.

Keywords Finiteness · Pseudofinite structures · Well-ordered sets · Strong minimality · Second-order logic

14.1 Beyond First-Order

This book is about mathematical structures and how logic is used to study them. The term that has been used most often is "first-order." By now we know what a structure is, but what is not clear at all is what is a property of a structure. There are many distinguishing features that can be chosen to describe a structure, but so far we have only paid attention to properties that are expressible by first-order formulas. The straightforward grammar of the language finds its counterpart in the geometry of definable relations, and gives rise to natural approaches to various classification problems. Moreover, this geometry is not a superficial construct imported from the world of logic into the world of mathematics. In the cases of the fields of real and complex numbers, it coincides with the geometries that are studied by algebraists and analysts. But first-order logic has its weaknesses.

Every structure with an infinite domain has a proper elementary extension. It follows that first-order properties cannot fully characterize infinite structures. For every structure with an infinite domain, there is a structure that has exactly the same first-order properties, but is not isomorphic to it. For a formal system in which structures can be fully characterized by their properties, one has to look beyond what is first-order expressible. It is a big subject and we will only touch it lightly, but before that let us see some more examples of properties that are first-order, and some that are not.

R. Kossak, *Mathematical Logic*, Springer Graduate Texts in Philosophy 3,
https://doi.org/10.1007/978-3-319-97298-5_14

14.1.1 Finiteness

Finiteness is a significant property; it is also a prime example of a property that is not first-order. There are sentences that force structure to be infinite. For example, one can say in one first-order statement that a binary relation is a linear ordering in which every element has an immediate successor, or that a linear ordering is dense. All structures equipped with such relations are infinite. However, first-order theories axiomatizing classical mathematical structure such as graphs, groups, rings, and fields, all have models with finite domains. But if a theory has only finite models, then there is an upper bound on their size. They cannot be arbitrarily large. If a theory has models with arbitrarily large finite domains, then it also must have infinite models, in fact it has models of arbitrarily large infinite cardinalities. Let s see why.

Let T be a theory[1] that has models with arbitrarily large finite domains. Then, T also has a model with an infinite domain. To prove it, onsider the theory T' consisting of T and the set of sentences φ_n, for $n = 1, 2, 3, \ldots$, where φ_n says that there are n different elements. Each finite fragment of T' contains only finitely many sentences φ_n; and since T has models with arbitrarily large finite domains, each finite fragment of T' has a model. By the compactness theorem, T' has a model. This model is a model of T, and its domain is infinite, because all sentences φ_n are true in it.

The fact that finiteness cannot be captured in a first-order way has interesting consequences. Let T be a theory that has models with arbitrarily large finite domains, and let T_{fin} be the set of sentences that are true in all finite models of T. Since all those finite models are models of T, T_{fin} contains T, and, as we saw above, T_{fin} must have infinite models. An infinite model of T_{fin} is called *pseudofinite* . Pseudofinite models are infinite, but retain some traces of finiteness. For example, let T be the theory of linear orderings. Every finite linear ordering has a smallest and a largest element. This can be expressed in a first-order way, hence all models of T_{fin} also must have a smallest and a largest element.

14.1.2 Minimality and Order-Minimality

Minimality is not a first-order property. In Chap. 12, we saw that $(\mathbb{N}, <)$ is minimal, and if $(\mathbb{N}^*, <)$ is one of its proper elementary extension, than $\mathbb{N}^*$ contains an isomorphic copy of $(\mathbb{Z}, <)$. If c is an element in that copy, then the set defined by the formula $x < c$ is neither finite nor cofinite in $\mathbb{N}^*$. If there were a sentence φ in the language with one relation symbol $<$, such that a structure is minimal if an only if φ is true in it, then φ would have to be true in $(\mathbb{N}, <)$ and in $(\mathbb{N}^*, <)$, and that can't be.

---

[1] A theory is a set of first-order sentences.

A structure is *strongly minimal* if it is minimal and every structure that has the same first-order theory is also minimal. The structure $(\mathbb{N}, <)$, and its dual $(\mathbb{N}, >)$, are rare examples of structures that are minimal but not strongly minimal. No other examples are known at the time I am writing these notes. It is easy to see that trivial structures are strongly minimal. It can be shown that the field of complex numbers is also strongly minimal.

There is no need for the notion of strongly order-minimal structure. Julia Knight, Anand Pillay, and Charles Steinhorn proved in 1986 that if $(M, <, \dots)$ is an order-minimal structure then so is every structure that has the same first-order theory [19]. In particular, if $(M, <, \dots)$ is order-minimal, then so is its every elementary extension. In this sense, order-minimality is a first-order property—the theory of the structure decides whether it is order-minimal or not, although there is no first-order theory in the language with just $<$ that decides that. If there were such a theory, it would be true in $(\mathbb{Q}, <)$, which is order-minimal, but then it would be also true in $(\mathbb{Q}, +, \cdot, <)$ (because it only speaks about $<$) which is as far from being order-minimal as possible.

14.2 Well-Ordered Sets and Second-Order Logic

Many properties of orderings are first-order. To have a first or last element, to be discrete, to be dense—these are all properties that can be expressed by a single first-order sentence about the ordering, but there is an important property of orderings that is not first-order. A linearly ordered set $(A, <)$ is *well-ordered* if every nonempty subset of A has least element. To see why this is not a first-order property, let us look again at $(\mathbb{N}, <)$ and its proper elementary extension $(\mathbb{N}^*, <)$. The natural numbers are well-ordered. Indeed, this is one of their most fundamental properties: every nonempty set of natural numbers has a least element. It is one of those clear and basic principles that do not require a proof. The integers $\mathbb{Z}$ are not well-ordered—the whole set $\mathbb{Z}$ does not have a least element, and hence every ordered set that contains a copy of $(\mathbb{Z}, <)$ is not well-ordered as well. Since $(\mathbb{N}, <)$ is well-ordered and its elementary extension $(\mathbb{N}^*, <)$ is not, it follows that the property of being well-ordered is not expressible by a first-order theory.

Let us now see how the property of being well-ordered can be expressed formally. We will do it in an extension of first-order logic, known as the *monadic second-order logic*. In the vocabulary of the monadic second-order logic, together with all the symbols of first-order logic, there are variables of the second-order sort X_1, X_2, ...; and there is an additional relation symbol $\in$, giving rise to atomic formulas of the form $x_i \in X_j$, for all i and j. In a structure, the first-order variables are interpreted in the usual way by individual elements of the domain, and the second order variables are interpreted by subsets of the domain. If x_i is interpreted by an element a and the variable X_j is interpreted by set of elements A, then, under this evaluation, the atomic formula $x_i \in X_j$ is true if an only if a is an element of the set A.

Here is a sentence that in a direct way expresses that $<$ is a well-ordering:

$$\forall X[\exists x\ (x \in X) \Longrightarrow (\exists x(x \in X \wedge \forall y(y \in X \Longrightarrow (y = x \vee x < y))))].$$

In the monadic second-order logic we can only quantify over subsets of the domain, as in the sentence above. In full second-order logic we can also quantify over arbitrary subsets of all Cartesian powers of the domain.

Second-order logic has a tremendous expressive power. Many infinite structures, such as $(\mathbb{N}, +, \times)$, can be completely characterized by their second-order properties. In fact, much of modern mathematics can be formalized using an axiom systems just about the second-order properties of $(\mathbb{N}, +, \times)$. Why then have we not used this logic to begin with? There is absolutely nothing wrong with the second-order logic but one has to use it with caution and sharper technical skills are necessary to do it well. When we approach a structure such as $(\mathbb{N}, +, \times)$ in a first-order fashion, we stand on a reasonably firm ground. The quantifiers $\exists$ and $\forall$ range over individual elements of the domain—the natural numbers. We have a good sense of what those elements are, and what it means that there is an element with some property, or that all elements have it. The moment we use $\exists X$ or $\forall X$, we need to pause to think a bit. What is the range of those quantifiers? They range over arbitrary sets of natural numbers, i.e. over the power set of $\mathbb{N}$. This is a much more elusive domain. To begin with, it is much larger than $\mathbb{N}$. The set of natural numbers is countable, its power set is not. All kind of questions arise. Some are of quite subtle character, for example, if a set A of natural numbers is defined by a second-order formula involving a quantifier $\forall X$, should A be considered in the range of this quantifier or not? The problem here is that if A allowed, then one can suspect some circularity—it seems that A is used in its own definition. Is this something that we want? There are other complications, due to the fact that many useful tools of first-order logic, such as the compactness theorem, are no longer available in second-order logic.

There is also the third-order logic, in which one can quantify sets of subsets of the domain, the fourth-order logic with quantifiers ranging over sets of sets of subsets of the domain. There is a whole hierarchy of logics that are used to express properties of sets in the hierarchy of sets that can be built over the domain of a structure. There are many other formal systems that extend first-order logic. There are systems in which the rules allow us to build infinite conjunctions and disjunctions of formulas, and there are systems in which one can also quantify infinite strings of variables. There are powerful logics obtained by adding new quantifiers. There is much more, but all those other logics always stand in comparison to first-order logic. They are either extensions or fragments of it and none of them has a model theory that is as well-developed as the model theory of first-order logic.

Exercises

Exercise 14.1 *Find a second order sentence that is true in a structure if and only if the structure is finite. Hint: Use the fact that a set* X *is infinite if and only if there is a one-to-one function* $f : X \longrightarrow X$*, such that for some* $b \in X$*,* $f(a) \neq b$ *for all* $a \in X$*.*

Exercise 14.2 * *Use the previous exercise to express minimality of structures as an axiom schema of second order logic. Hint: For each formula* $\varphi(x)$ *of the language of the structure, include axioms of the form "either the set of* x *such that* $\varphi(x)$ *is finite, or the set of* x *such that* $\neg\varphi(x)$ *is finite."*

Exercise 14.3 * *Modify your answer to the previous exercise to express order-minimality of structures as an axiom schema of second order logic.*

Exercise 14.4 * *Suppose that all second-order sentences that are true in* $(\mathbb{N}, +, \cdot)$ *are also true in* $(M, \oplus, \odot)$*. Prove that* $(M, \oplus, \odot)$ *is isomorphic to* $(\mathbb{N}, +, \cdot)$*. Hint: Let* 0_M *and* 1_M *be the first two elements in the ordering of* M*. Define* $f : \mathbb{N} \longrightarrow M$ *by induction:* $f(0) = 0_M$*, and* $f(n+1) = f(n) \oplus 1_M$*. Let* $N = \{f(n) : n \in \mathbb{N}\}$*. Show that* $(N, \oplus, \odot)$ *is isomorphic to* $(\mathbb{N}, +, \cdot)$*, and that* $N = M$*.*

Chapter 15
Symmetries and Logical Visibility One More Time

Abstract My aim in this book was to explain the concept of mathematical structure, and to show examples of techniques that are used to study them. It would be hard to do it honestly without introducing some elements of logic and set theory. In a textbook, the line of thought may sometimes get lost in technical details. Now, when all necessary material has been covered, I can give a summary and a brief description of what this book is about.

15.1 Structures

What is a structure? It is a set, called the domain, with a set of relations on it. Any set can serve as a domain of a structure. In the logic approach to structures, the first step is to forget about what the elements of the domain are made of, and to consider only how the elements and their finite sequences (ordered pairs, triples, etc.) are related by the relations of the structure. To know a structure defined this way, means to understand the geometry of its parametrically definable sets. We are not only interested in the subsets of the domain, but also in the whole multidimensional spectrum of the definable subsets of all of Cartesian powers of the domain.

Where do all the sets come from? This is a difficult question. In modern mathematics we rely on the axiomatic method. The axioms determine what we know about the universe of sets. This is the reason for the detour through the axioms of the Zermelo-Fraenkel set theory in Chap. 6.

Once the status of sets and relations gets clarified, the question is what do we want to know about a mathematical structure. What is there to see in it? Here, for a partial answer we resorted to first-order logic. What we "see" in a structure is what can be defined from its relations by means of logic, hence we can talk about a sort of *logical seeing*.

Let X be a domain of a structure. For each $n > 0$, the parametrically definable subsets of X^n form a Boolean algebra, which means that if A and B are parametrically definable subsets of X^n, then so are their intersection $A \cap B$, their union $A \cup B$, and their complements $\bar{A}$ and $\bar{B}$, and those set operations correspond, respectively, to forming conjunctions, disjunctions, and negations of the defining

R. Kossak, *Mathematical Logic*, Springer Graduate Texts in Philosophy 3,
https://doi.org/10.1007/978-3-319-97298-5_15

formulas. Those are the basic sets that we see through the eyes of logic, but the formalism of first-order logic allows us to describe more sets. Taking Boolean combinations of formulas with different sequences of variables results in definable sets in higher dimensions. For example if $\varphi(x_1, x_2)$ and $\psi(x_3)$ are formulas with free variables x_1, x_2, and x_3 respectively, then $\varphi(x_1, x_2)$ defines a subset of X^2, $\psi(x_3)$ defines a subset of X, and $\varphi(x_1, x_2) \wedge \psi(x_3)$ defines a subset of X^3. Finally, quantification corresponds to projections from higher dimensions to lower.

In the case of tame structures, the geometry of one-dimensional definable sets to large extent determines the geometry in higher dimensions, and the general theory of such structures leads to descriptions and classifications of all parametrically definable sets. We do understand those structures well. We see them. In this book, we could not get into more advanced details of the model theory of tame structures, but we could see some elements of the theory and the role that first-order logic plays in it.[1]

Wild structures are just wild, and there is no hope for any general theory of all of their parametrically definable sets. The first-order analysis of those wild structures reveals an enormous complexity in the geometry of the definable sets. From the basic relations, using mechanical rules for the syntax of first-order logic one can generate a whole hierarchy of mathematical objects with each level more complex than the previous one.

In both cases, tame and wild, symmetries are an important tool in the analysis of structures. Much information about a structure can be gained by the study of the subsets of its domain that are invariant under symmetries. Think of corner points of a square, or the central point of a star-shaped graph. If a structure has symmetries, we can learn much from describing and classifying them, but many structures do not have symmetries, and then it helps to consider elementary extensions instead. Every structure with an infinite domain has an elementary extension to a larger structure. This combined with the power of set-theoretic methods, allows us to construct structures with interesting properties. In particular, every structure with an infinite domain can be extended to a structure with many symmetries.

There are many attractive examples we could have discussed, but we concentrated on the classical number structures. It is quite fascinating to see the historical development of the number system, and then to see how the corresponding number structures grew from very wild, to completely tame. It took many pages to tell that story. Now we can tell it again briefly.

15.2 The Natural Numbers

The German mathematician Leopold Kronecker famously said that "God made the integers, all else is the work of man." We outlined that work in great detail. We built everything from the natural numbers, but we did not assume that God gave them to us, we built them from scratch ourselves.

[1]For a technical exposition see [36].

The domain $\mathbb{N}$ can be defined in many ways. In set theory, following John von Neumann, we define them by declaring that the empty set $\varnothing$ stands for 0, then $\{\varnothing\}$ is 1, $\{\varnothing, \{\varnothing\}\}$ is 2, $\{\varnothing, \{\varnothing\}, \{\varnothing, \{\varnothing\}\}\}$ is 3, and so on. We discussed this in Chap. 6. Defining natural numbers this way has its advantages, but there is nothing sacred about this definition. Before von Neumann, Ernst Zermelo defined natural numbers to be: $\varnothing, \{\varnothing\}, \{\{\varnothing\}\}, \{\{\{\varnothing\}\}\}\ldots$ and that is fine too. The whole point here was not to start with some assumed common conception of what the natural numbers should be, but to ground the notion in a more rudimentary set-theoretic background.

The natural numbers form a set and this set as such is a trivial structure, not different from any other countable infinite set. It becomes a fascinating structure once it is equipped with addition and multiplication. Von Neumann's approach allows us to give straightforward set-theoretic definitions of both operations, and thus the structure $(\mathbb{N}, +, \cdot)$ is born. After all this effort, we can forget about what the natural numbers are made of. We have a set and a set of two relations on it, and this is all we need for further analysis.

The natural numbers are linearly ordered, but there is no need to include the ordering relation since it is definable in $(\mathbb{N}, +, \cdot)$. Since the defining formula $\neg(x = y) \wedge \exists z(x + z = y)$ does not involve multiplication, the relation $x < y$ is already definable in $(\mathbb{N}, +)$. Using the ordering, it is easy to see that every natural number is definable. This is a fundamental feature of the natural numbers. There is an old mathematical joke about a theorem that says "Every natural number is interesting." Indeed, 0 is interesting because it is the least number; 1 is interesting because it is the least positive number; 2 is the least even number; 3 is the least odd prime number; 4 is the least even square; and so on. To prove the theorem, suppose that there is a natural number that is not interesting. Then there must be a least such number m. So all numbers smaller than m are interesting, but m is not. That is a curious property, and it clearly makes m interesting. So m is interesting after all, and this contradiction finishes the proof of the theorem.

Since every natural number is definable in $(\mathbb{N}, <)$ without parameters, this structure is rigid, it has no nontrivial symmetries. In order to show that $(\mathbb{N}, <)$ is minimal, we used symmetries of its elementary extension $(\mathbb{N}^*, <)$. Even though the successor function $S(x) = x + 1$ is definable in $(\mathbb{N}, <)$, the whole addition relation is not. If we could define addition in $(\mathbb{N}, <)$, it would follow that $(\mathbb{N}, +)$ is minimal, but we know that it is not.[2]

The structure $(\mathbb{N}, +)$ is not minimal, but still quite tame. By theorem of Ginsburg and Spanier, proved in 1966, every set of natural numbers definable in $(\mathbb{N}, +)$ is ultimately periodic, which means that for every such set X there are numbers m and p such that for any n, if $m < n$ then n is a member of X if and only if $n + p$ is. After an unpredictable initial segment, every definable set becomes quite well-behaved. This rules out a possibility of defining sets such as the set of all square numbers, or the set of primes. None of them is ultimately periodic. When we expand $(\mathbb{N}, +)$ by adding multiplication, it gets wild. $(\mathbb{N}, +, \cdot)$ is the ultimate wild structure. Any other structure that contains a definable isomorphic copy of $(\mathbb{N}, +, \cdot)$ is also wild.

[2]For example, the formula $\exists y[y + y = x]$ defines the set of even numbers (which is neither finite nor cofinite).

The multiplicative structure on natural numbers $(\mathbb{N}, \cdot)$ is also not minimal. In the language of multiplication one can define the set of square numbers (which is neither finite nor cofinite). The set of primes is also definable. It is interesting though that the ordering of the natural numbers is not definable. This follows from the fact that $(\mathbb{N}, \cdot)$ has nontrivial automorphisms. If p and q are prime numbers, then the function that permutes p and q and fixes all other prime numbers can be extended to a symmetry of $(\mathbb{N}, \cdot)$ (this is not obvious, see hints in the exercise section). Hence, in $(\mathbb{N}, \cdot)$ all prime numbers share the same type. Since 2 is prime, and all other primes are odd, it also follows, that the set of even numbers is not definable in $(\mathbb{N}, \cdot)$.

The results about definability in $(\mathbb{N}, +)$ and $(\mathbb{N}, \times)$ imply that addition of natural numbers is not definable from multiplication, and that multiplication is not definable from addition. This is an intriguing corollary. Even though addition of natural numbers *determines* their multiplication, first-order logic does not capture it.

15.3 The Integers

The set of integers $\mathbb{Z}$ is obtained by appending negative opposites of all natural number to $\mathbb{N}$. To make this "appending" precise, in Chap. 4, we defined the integers in a special way to show how the extended structure $(\mathbb{Z}, +, \cdot)$ can be defined in $(\mathbb{N}, +, \cdot)$. This showed how, from the logical point of view, the larger structure is no more complex than the one it contains. It also turned out that the larger structure is not less complex, because $\mathbb{N}$ is definable in $(\mathbb{Z}, +, \cdot)$, for example by the following formula $\varphi(x)$ in which x is the only free variable

$$\exists x_1 \exists x_2 \exists x_3 \exists x_4 [x = x_1^2 + x_2^2 + x_3^2 + x_4^2].$$

This works because of a theorem of Joseph-Louis Lagrange, who proved in 1770 that every positive integer is the sum of four squares. Notice that the formula involves both $+$ and $\cdot$. Equipped with $\varphi(x)$ we can also define the ordering of the integers, since for all integers x and y, $x < y$ if and only if $y + (-x)$ is positive. Hence, the relation $x < y$ is defined by the formula

$$\exists v \exists w [x + v = 0 \wedge w = y + v \wedge \varphi(w)].$$

This definition is decidedly more involved that the one that defines the ordering of the natural numbers in $(\mathbb{N}, +)$. One reason is that the use of multiplication in $\varphi(x)$ is not accidental. The ordering of the integers is not definable in $(\mathbb{Z}, +)$. The function $f(x) = -x$ is a symmetry of $(\mathbb{Z}, +)$, and consequently the type of any integer is the same as the type of its opposite. Addition alone cannot decide whether $0 < 1$ or $0 < -1$, because $f(0) = 0$, and $f(1) = -1$, hence $(0, 1)$ and $(0, -1)$ have the same first-order properties in $(\mathbb{Z}, +)$.

The argument above shows that 1 is not definable without parameters in $(\mathbb{Z}, +)$. It is definable in $(\mathbb{Z}, +, \cdot)$ as the only number x other than 0, such that $x \cdot x = x$. Since 1 and the ordering are definable in $(\mathbb{Z}, +, \cdot)$, it follows that every integer is definable; hence $(\mathbb{Z}, +, \cdot)$ is rigid.

15.4 The Rationals

The set of rational numbers $\mathbb{Q}$ together with addition and multiplication is first-order definable in $(\mathbb{Z}, +, \cdot)$, hence it is also first-order definable in $(\mathbb{N}, +, \cdot)$. Since 0 and 1 are definable in $(\mathbb{Q}, +, \cdot)$ by the same formulas that defined them in $(\mathbb{Z}, +, \cdot)$, one can show that all rational numbers are definable; hence $(\mathbb{Q}, +, \cdot)$ is rigid. However, $(\mathbb{Q}, +)$ has more symmetries than $(\mathbb{Z}, +)$. While $f(x) = -x$ is the only nontrivial symmetry of $(\mathbb{Z}, +)$, $(\mathbb{Q}, +)$ has many more. In is not difficult to see, and we leave it as an exercise, that for every positive rational number p, the function $f_p(x) = p \cdot x$ is a symmetry.

Lagrange's theorem also holds for the rationals: each non-negative rational number is the sum of fours squares of rational numbers. Hence, the ordering of the rationals is definable in $(\mathbb{Q}, +, \cdot)$.

15.5 The Reals

With the introduction of the real numbers we enter a completely new territory. The set of real numbers is uncountable, so it cannot be definably interpreted in $(\mathbb{N}, +, \cdot)$. It is too large. To construct it we needed to appeal to set theory. Real numbers are complex objects as individuals, but the structure of the field of real numbers is much less complex than that of the rationals. It takes much less effort to define the ordering: the relation $x < y$ for the real numbers is defined by the formula $\neg(x = y) \wedge \exists z[x + z \cdot z = y]$, and this is due to the fact that every positive real number has a square root.

The field $(\mathbb{R}, +, \cdot)$ is order-minimal, and, as a consequence, neither $\mathbb{N}$, nor $\mathbb{Z}$, nor $\mathbb{Q}$ are definable in it. The rational numbers are not definable in $(\mathbb{R}, +, \cdot)$ as a set, but each rational number is defined by the same definition that defines it in $(\mathbb{Q}, +, \cdot)$. For each real number r, the type of r over $(\mathbb{R}, +, \cdot)$ includes the formulas $p < x$, and $x < q$, for all rational p and q such that $p < r$, and $r < q$. In fact, r is the only real number making all those formulas true. It follows that $(\mathbb{R}, +, \cdot)$ is rigid. No rational number can be moved by a symmetry because they are all definable without parameters, and no irrational number can be moved, because it is firmly trapped between two sets of rational numbers.

There are more definable reals than just the rationals. Each polynomial equation in one unknown with integer coefficients is either an identity, or has a finite number of solutions. Since the ordering is definable, for a given equation $p(x) = 0$ that has finitely many solutions, its smallest solution is definable, the next smallest solution is definable, and so on. Hence all real solutions of such equations are definable. The numbers that are members of finite solution sets of polynomial equations with integer coefficients are called *algebraic*.

The set of algebraic numbers includes many irrational numbers such as $\sqrt{2}$ or the golden ratio $\frac{1+\sqrt{5}}{2}$, and it is closed under addition and multiplication, and additive and multiplicative inverses, i.e. if a and b are algebraic then so are $a + b$, $a \cdot b$, and $-a$, and $\frac{1}{a}$ (if $a \neq 0$). Since the collection of all polynomial equations with integer coefficients is countable, and each nontrivial equation has only finitely many solutions, the set of real algebraic numbers is countable. The set of real numbers is uncountable, hence there are many real numbers that are not algebraic. Those numbers are called *transcendental*. Many important mathematical constants, such as π and Euler's constant e, are transcendental, but in each case it takes an effort to prove it. It can be shown that the real algebraic numbers are the only numbers that are definable in $(\mathbb{R}, +, \cdot)$. It follows that there are no first-order definitions of π and e, even though they are uniquely determined by their first-order types (i.e. their Dedekind cuts). Transcendentals are not logically visible in $(\mathbb{R}, +, \cdot)$.

We can also say a bit about parametrically definable sets of real numbers. It follows from order-minimality that if a set A of real numbers is infinite, and is parametrically definable in $(\mathbb{R}, +, \cdot)$, then it must contain an interval, and that implies that it is not only infinite, but is in fact uncountable. Every interval of the real line has the same cardinality as the whole set of real numbers. It is of power continuum. Hence, every parametrically definable in $(\mathbb{R}, +, \cdot)$ set of real numbers is either finite (small), or of the power continuum (very large). No formula defines a set of cardinality $\aleph_0$.

15.6 The Complex Numbers

The field of the complex numbers $(\mathbb{C}, +, \cdot)$ is the ultimate number structure. It has an interesting history and remarkable applications. Much more space would be needed to do justice to all that. We will just scrape the surface.

The field $(\mathbb{C}, +, \cdot)$ is minimal. Every parametrically definable in $(\mathbb{C}, +, \cdot)$ subset of $\mathbb{C}$ is either finite or cofinite. Complex numbers that are elements of finite sets definable without parameters are called *algebraic*. Algebraic numbers are solutions of polynomial equations with integer coefficients. All real algebraic numbers, as defined in the previous section, are algebraic in the complex sense, and there are many more complex algebraic numbers that are not real. Prime examples the imaginary unit $i = (0, 1)$, and its additive inverse $-i = (0, -1)$.[3] Since there are only countably many definitions, and the union of countable many finite sets in countable, it follows that there are many non-algebraic complex numbers, and those numbers are also called transcendental. All real transcendental numbers are also transcendental as complex numbers, but unlike in the real field, where each number has its own unique type, in the complex field all transcendental numbers share the same type. There is only one type of a transcendental number in $(\mathbb{C}, +, \cdot)$! Even more is true, if z_1 and z_2 are transcendental, then there is a symmetry f of $(\mathbb{C}, +, \cdot)$ such that $f(z_1) = z_2$. In particular, there is a symmetry f such that $f(\pi) = e$. While the real field is rigid, the complex field has lots and lots of symmetries.

[3] i and $-i$ are the only solutions of $x^2 + 1 = 0$.

Here is one example of a nontrivial symmetry. The function that maps (a, b) to $(a, -b)$ is called *conjugation*. It is not difficult to check that conjugation is a symmetry of $(\mathbb{C}, +, \cdot)$. Under this symmetry, the image of i is $-i$, hence those two algebraic numbers have the same type. The numbers i and $-i$ have exactly the same first-order properties in $(\mathbb{C}, +, \cdot)$, they are indiscernible.

Conjugation maps every real number $r = (r, 0)$ to itself. All real numbers are fixed under conjugation. It follows that the graph of conjugation, i.e. the set G of all conjugate pairs of complex numbers $((a, b), (a, -b))$ is not a definable subset of $\mathbb{C}^2$. If G were definable by some formula $\varphi(x, y)$, then the set of real numbers would be defined by the formula $\varphi(x, x)$, but we know that it cannot be defined in $(\mathbb{C}, +, \cdot)$, because it is neither finite nor cofinite in $\mathbb{C}$.

Using conjugation we can also show that there cannot be any definable linear ordering of the complex numbers. For each linear ordering $<$, either $-i < i$ or $i < -i$, but conjugation swaps i with $-i$; hence, if the ordering were definable, it would follow that $-i < i$ if and only if $i < -i$, but that can't be.

All those results illustrate that there is very little one can see in $(\mathbb{C}+, \cdot)$ with the eyes of logic. This is certainly a weakness, since there is much going on there. One of the most well-known modern mathematical images is probably the *Mandelbrot set*. Many videos showing zooms deep into the boundary of the Mandelbrot set are available on youtube. They show the immense complexity of the set that is defined in a surprisingly simple way. For any complex number c, one starts with 0, and then iterates the function $z \mapsto z^2 + c$. This defines a sequence of numbers that either converges to 0, or stays in a bounded region of $\mathbb{C}$, or moves farther and farther away from 0 "escaping to infinity." The Mandelbrot set is the set of those numbers c for which the sequence does not escape to infinity.

The procedure used to define the Mandelbrot set involves addition and multiplication of complex numbers but it cannot be written in first-order fashion over $(\mathbb{C}+, \cdot)$, because the Mandelbrot set is neither finite nor cofinite. This means that there are concepts used in the definition that cannot be formalized in first-order logic in the language of the field of complex numbers. It is a subtle issue, so let us take a closer look.

The definition refers to distances of complex numbers from 0. This distance of a point (a, b) from 0 is perfectly well-defined in $(\mathbb{R}, +, \cdot)$; it is $\sqrt{a^2 + b^2}$. The function $f(a, b) = \sqrt{a^2 + b^2}$ is definable in $(\mathbb{R}, +, \cdot)$, but it cannot be definable over $(\mathbb{C}, +, \cdot)$ because its range is neither finite nor cofinite.

One could ask if the Mandelbrot set is second-order definable. In second-order logic one can define much. For example, let us take a look at the following second-order formula $\Phi(x)$.

$$\forall X[(0 \in X \wedge \forall z(z \in X \Longrightarrow z + 1 \in X) \Longrightarrow x \in X].$$

$\Phi(x)$ says that x belongs to any set that contains 0 and is closed under successor. In other words, x is in the smallest X with that property. Such a smallest set exists, and it is exactly the set of natural numbers $\mathbb{N}$. The field $(\mathbb{C}+, \cdot)$ is minimal, hence one can not define $\mathbb{N}$ in it by a first-order formula, but we just showed that it can be done by a second-order one. Since $\mathbb{N}$ is second-order definable over $(\mathbb{C}, +, \times)$ a whole copy of $(\mathbb{N}, +, \cdot)$ is second-order definable as well, and that means that the second-order structure of the complex field is very complex indeed. Still, the Mandelbrot set is not second-order definable. That is because symmetries preserve not only the first-order definable properties, but all second-order properties as well. Suppose $M(x)$ were a second-order formula defining the Mandelbrot set. The Mandelbrot set contains small transcendental numbers, for example $\frac{1}{\pi}$; hence $M(\frac{1}{\pi})$ holds in $(\mathbb{C}, +, \cdot)$. But that implies that every transcendental number r is in the Mandelbrot set, since for every such number there is a symmetry f of $(\mathbb{C}, +, \times)$ such that $f(\frac{1}{\pi}) = r$. That is a contradiction, since the Mandelbrot set is bounded, and there are unboundedly many transcendental numbers, for example π is too large to be in the Mandelbrot set.

The discussion above is not intended to show that logical formalism is so deficient that it cannot capture something that is naturally defined in a mathematical language. The procedure that defines the Mandelbrot set can be easily converted to a second-order formula that defines that set over $(\mathbb{R}, +, \cdot)$. The reader who was patient enough to get to this point, may be ready to try to write such a formula.

Exercises

Exercise 15.1 *The Fundamental Theorem of Arithmetic says that every natural number can be uniquely written as a product of powers of prime numbers (in increasing order of primes). For example* $12 = 2^2 \cdot 3$, *and* $300 = 2^2 \cdot 3 \cdot 5^2$. *Hence, every number can be written in the form* $2^k \cdot 3^l \cdot m$ *for some natural number m. If* 2 *or* 3 *are missing in the representation, then the corresponding k or l are equal to* 0. *Consider the function* $f : \mathbb{N} \longrightarrow \mathbb{N}$, *defined by* $f(2^k \cdot 3^l \cdot m) = 3^k \cdot 2^l \cdot m$. *Show that* f *is a symmetry of* $(\mathbb{N}, \cdot)$.

Exercise 15.2 *Use the previous exercise to show that the ordering of the natural numbers is not definable in* $(\mathbb{N}, \cdot)$.

Exercise 15.3 *Show that 0 is the only integer that is definable without parameters in* $(\mathbb{Z}, +)$. *Hint: The function* $f : \mathbb{Z} \longrightarrow \mathbb{Z}$ *defined by* $f(x) = -x$ *is a symmetry of* $(\mathbb{Z}, +)$.

Exercise 15.4 *Show that every rational number is definable without parameters in* $(\mathbb{Q}, +, \cdot)$. *Hint: See Exercise* 8.1.

Exercise 15.5 *Show that for every rational number p other than* 0, *the function* $f_p(x) = p \cdot x$ *is a symmetry of* $(\mathbb{Q}, +)$.

Exercise 15.6 *Every fraction* $\frac{p}{q}$ *can be written as* $\frac{p \cdot q}{q^2}$. *Apply this and Lagrange's theorem to prove that every positive rational number is equal to the sum of fours squares of rational numbers. This shows that the ordering of the rational numbers is definable in* $(\mathbb{Q}, +, \cdot)$.

Exercise 15.7 *The two solutions of the equation* $x^2 - x - 1 = 0$ *are* $\frac{1+\sqrt{5}}{2}$ *and* $\frac{1-\sqrt{5}}{2}$. *Write first-order formulas defining each of these numbers in* $(\mathbb{R}, +, \cdot)$.

Exercise 15.8 *Show that the function* $f : \mathbb{C} \longrightarrow \mathbb{C}$ *defined by* $f(a, b) = (a, -b)$ *is a symmetry of* $(\mathbb{C}, +, \cdot)$.

Exercise 15.9 * *Find a second-order definition of the Mandelbrot set over* $(\mathbb{R}, +, \cdot)$.

Chapter 16
Suggestions for Further Reading

Needles to say, there is a vast literature on model theory of first-order logic and its applications. Some references have already been given throughout the text. I will repeat some of them and will add other recommendations.

Much can be learned about any subject by studying its history. Mathematical logic has a well documented and quite intriguing past. For a very comprehensive account, I recommend Paolo Mancosu's *The Adventure of Reason. Interplay between philosophy of mathematics and mathematical logic: 1900–1940* [20]. For a shorter discussion of issues around the origin and the history of definitions by abstraction and the concept of infinity, see Mancosu's *Abstraction and Infinity* [21].

Stanford Encyclopedia of Philosophy has informative entries on most topics that have been discussed here. In particular, see the entries on Zermelo's axioms of set theory [9] and first-order model theory [15]. For more on recent developments in set theory and model theory see [2] and [14].

From its inception, set theory has always been a tantalizing subject not only for the philosophy of mathematics, but also for philosophy in general. In his review of Alain Badiou's *Being and Event* [1], Peter Dews writes:

> All we can say ontologically about the world, Badiou contends, is that it consists of multiplicities of multiplicities, which can themselves be further decomposed, without end. The only process which endows any structure on this world is the process which Badiou calls 'counting-as-one', or—in more explicitly mathematical terms—the treatment of elements of whatever kind, and however disparate, as belonging to a set. We must not imagine, prior to this counting, any ultimate ground of reality [12].

In contrast, for many years model theory stayed in a shadow. This seems to be changing now. Recently published books by John Baldwin [3], Tim Button and Sean Walsh [4], and Fernando Zalamea [39], take the philosophy of model theory as their subjects.

For more about more recent mathematical advances in model theory, I recommend Boris Zilber's chapters in [25]. That book is also highly recommended for its unique vision of mathematical logic.

R. Kossak, *Mathematical Logic*, Springer Graduate Texts in Philosophy 3,
https://doi.org/10.1007/978-3-319-97298-5_16

A full history of model theory still waits to be written, but a very good introduction to it is Wilfrid Hodges' appendix in [4].

It is interesting to compare the contemporary view of the discipline with its descriptions in the past. Andrzej Mostowski's *Thirty years of foundational studies: lectures on the development of mathematical logic and the study of the foundations of mathematics in 1930–1964* [26], reprinted in [27], gives a very thorough outline of mathematical logic up to the 1960s, including a chapter on model theory.

In Part I in this book, much attention was given to the development of the basic number systems in the logical framework. I roughly followed a much more detailed presentation given by Craig Smoryński in [32], and the reader is referred to this excellent book for all details that were skipped here. For a briefer, but also very informative description of how large parts of mathematics, including portions of analysis and algebra, can be formally developed starting from the standard model of arithmetic $(\mathbb{N}, +, \cdot)$, see introductory chapters in John Stillwell's [33]. Stillwell's book is devoted to a relatively new, very attractive direction in foundational studies known as *Reverse Mathematics*.

Finally, I recommend two biographies by Constance Reid: *Hilbert* [30] (that book has had many editions), and *Julia, a Life in Mathematics* [29] about the life and work of her sister Julia Robinson and the history of the solution of Hilbert's 10th problem.

Appendix A
Proofs

A.1 Irrationality of $\sqrt{2}$

There are many different proofs of irrationality of $\sqrt{2}$. The one that I will present here is due to the American mathematician Stanley Tennenbaum (1927–2005). It is a beautiful proof.

We need some preparation. Consider a unit square pictured below. The length of each side is 1. Let d be the length of the diagonal.

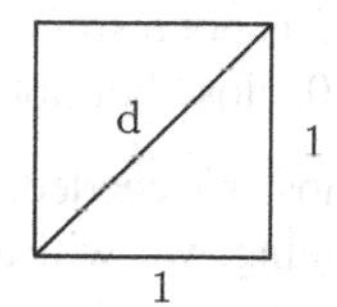

Now consider a d by d square positioned as in the picture below. The dashed lines show a right triangle. The smaller square can be cut into two such triangles, the larger one into four, which shows that the area of the larger square is 2. The larger square is twice larger than the smaller.

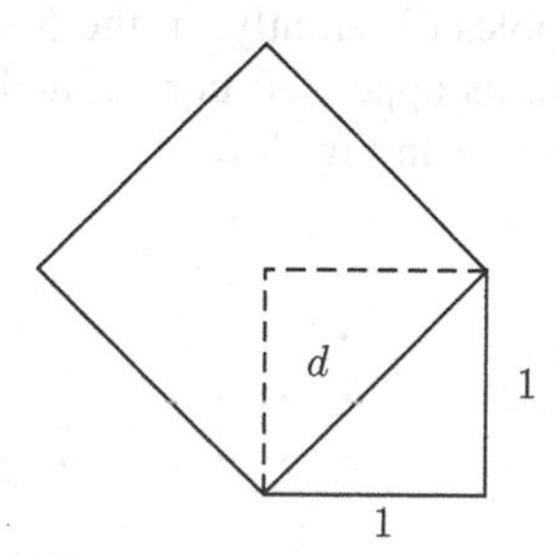

R. Kossak, *Mathematical Logic*, Springer Graduate Texts in Philosophy 3,
https://doi.org/10.1007/978-3-319-97298-5

Because the area of the larger square is 2, the length of its side d must be such that $d^2 = 2$. Hence, d is the square root of 2. The picture shows that d is a number between 1 and 2. One can ask if by choosing smaller units of length, one could measure the side of the square with exactly m such units, and the diagonal by exactly n units, where m and n are natural numbers.

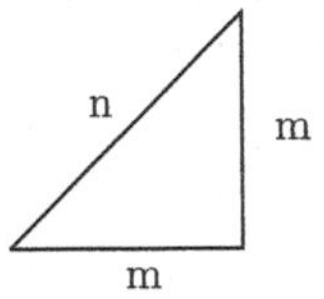

If we could find such m and n, it would follow that n^2 is equal to twice m^2, i.e. $2m^2 = n^2$. Then, by dividing both sides by m^2 we would obtain that $2 = (\frac{n}{m})^2$, and that in turn would show that $\sqrt{2} = \frac{n}{m}$, proving that $\sqrt{2}$ is a rational number. Tennenbaum's proof shows that there can be no such m and n. The proof begins now.

Suppose that there is a number m, such that $2m^2$ is a square, i.e. $m^2 + m^2 = n^2$ for some natural number n. Let's see what this m could be.

- Could $m = 1$? No, $1^2 + 1^2 = 2$ is not a square.
- Could $m = 2$? No, $2^2 + 2^2 = 8$ is not a square.
- Could $m = 3$? No, $3^2 + 3^2 = 18$ is not a square.
- Could $m = 4$? No, $4^2 + 4^2 = 32$ is not a square.
- Could $m = 5$? No, $5^2 + 5^2 = 50$, close but not a square.

For $m = 2$ and $m = 5$ we almost succeeded, we missed a square by 1. It is encouraging, but instead of continuing, we will assume that we have found an m that works, and we will examine consequences.

In preparation, let us consider a similar case. It is well know that the sum of squares can be a square. It happens infinitely often. We will illustrate what is going on using pebbles in square boxes. For example, $3^2+4^2 = 5^2$. Fig. A.1, it shows how 9 pebbles and 16 pebbles are first arranged in two square boxes, and then rearranged into five rows with five pebbles in each row in a 5 by 5 square.

Now let rearrange the pebbles differently. In the 5 by 5 square let us place the 9 pebbles to form a square in its upper left corner, and then 16 pebbles to form a square in the lower right corner as in Fig. A.2.

Fig. A.1 $3^2 + 4^2 = 5^2$

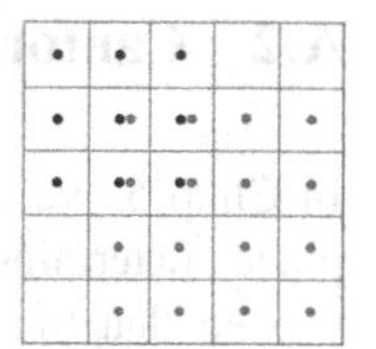

Fig. A.2 Squares rearranged

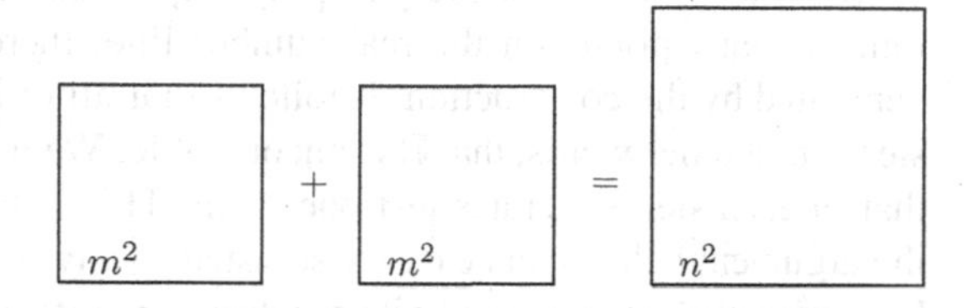

Fig. A.3 $m^2 + m^2 = n^2$

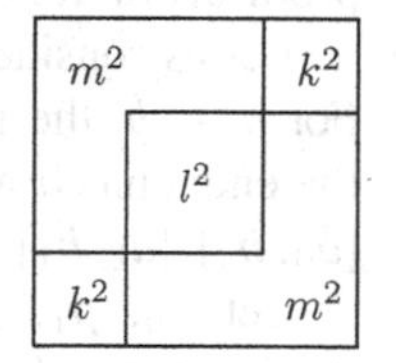

Fig. A.4 Rearranging $m^2 + m^2$

The squares overlap, and the number of doubly occupied slots is equal to the number of empty slots in the other two corners of the big square. It is clear that it is the case by inspecting the picture. For the rest of the proof, keep in mind that when the squares arranged the same fashion as in Fig. A.2, for each doubly occupied slot, there must be one that is left unoccupied.

Now we come back to Tennebaum's proof. Think about what would happen if instead of adding two squares of different sizes, we could add m^2 to m^2 and obtain a bigger square n^2 (Fig. A.3).

The picture will be similar to the one in Fig. A.2, but now since the smaller squares are of the same size, the picture is symmetric, as in Fig. A.4. The region over which the smaller squares overlap must consist of the same number of slots, as the two empty areas. Hence, if l is the length of each side of the square in the middle, and k is the length of each sides of the two smaller squares, then $k^2 + k^2 = l^2$.

Now comes the punch line. If there were a whole number m such that $m^2 + m^2$ is a square then, there would be a smallest such number m. Suppose that m is that smallest number. Then, the argument above shows that there is a smaller natural number k such that $k^2 + k^2 = l^2$, so $k^2 + k^2$ is a square, hence m could not have been the smallest such number, and that is a contradiction and it completes the proof. $\sqrt{2}$ is not rational.

A.2 Cantor's Theorem

In Chap. 5, we saw that the real line $\mathbb{R}$ is made of points that are ideal geometric places determined by the set of *all* Dedekind cuts in the rational numbers. Cantor proved that the collection of all such cuts cannot be obtained in a step-by-step process. In this section we will examine Cantor's argument.

We will show that for every step-by-step construction that at each step generates finitely many points on the real number line, there is always a point that is not generated by the construction. It follows that all of $\mathbb{R}$ cannot be generated step-by-step, or, in other words, that $\mathbb{R}$ is uncountable. We will consider an arbitrary process that in each step generates just one point. This is for notational convenience only, the argument below can be easily adjusted to any other step-by-step construction as long as at each step only a finite number of points in generated. When you read the proof below, it helps to draw pictures.

Let us consider a step-by-step process that generates points on the number line. For $a < b$, the interval $[a, b]$ is the set of all points between a and b, including the endpoints a and b. We will use the process to construct a sequence of intervals $[a_0, b_0]$, $[a_1, b_1]$, $[a_2, b_2]$, ..., each next interval contained in the previous one.

Let p_0, p_1, p_2,... be the sequence of points that our step-by-step process produces one by one. To begin our construction, we take any interval $[a_0, b_0]$ with rational endpoints, and select a rational point q that is different from p_0 and $a_0 < q < b_0$. Since p_0 is different from q, it can only be in one of the intervals $[a_0, q]$ and $[q, b_0]$. If it is not in the first interval, we let $[a_1, b_1]$, be $[a_0, q]$; otherwise we let it be $[q, b_0]$. So p_0 is not in $[a_1, b_1]$. In the second step, we select another rational q, this time different from p_1, such that $a_1 < q < b_1$. We consider intervals $[a_1, q]$ and $[q, b_1]$ and select as $[a_2, b_2]$ the one which does not contain p_1. So $[a_2, b_2]$ contains neither p_0 nor p_1. We continue in this fashion to obtain a nested sequence of nonempty intervals $[a_n, b_n]$ such that for each n, $[a_n, b_n]$ does not contain any of the points $p_0, p_1, \dots, p_{n-1}$.

Now we add an additional twist to the selection of the points a_n and b_n: we want the distance between them to be smaller than $\frac{1}{n}$. This can be achieved by moving a_n closer to b_n, if a_n and b_n do not satisfy this additional condition already. For such modified sequence of intervals, let $D = \{p : p \in \mathbb{Q} \wedge \exists n\ (p \leq a_n)\}$. Then, D is a Dedekind cut that corresponds to a real number d. Since d belongs to all intervals $[a_n, b_n]$, it must be different from all points p_n, and this finishes the proof.

An interesting feature of this proof is that if the procedure that generates the sequence p_0, p_1, p_2,... is not just "given," but actually is an effective computational procedure, then from the proof we can extract an algorithm to compute the digits in the decimal expansion of the number d. In particular, since there is an effective procedure that lists all real algebraic numbers, we get an algorithm to construct a non-algebraic number.

A.3 Theories of Structures with Finite Domains

In Chap. 11, we noted that structures with finite domains that share the same first-order theories must be isomorphic. This may not be that surprising if one thinks of a structure with a finite number of relations. It seems reasonable that a finite amount of information suffices to describe what those relations on a finite domain look like. The proof that is given below is interesting because it does not attempt to show how a finite structure can be completely described by its first-order properties. It is a proof by contradiction, and it works for an arbitrary number of relations.

Theorem A.1 *Let $\mathfrak{A}$ and $\mathfrak{B}$ be structures with finite domains for the same first-order language. If for every first-order sentence φ, φ is true in $\mathfrak{A}$ if and only if it is true in $\mathfrak{B}$, then $\mathfrak{A}$ and $\mathfrak{B}$ are isomorphic.*

Proof Suppose, to the contrary, that $\mathfrak{A}$ and $\mathfrak{B}$ satisfy the assumptions of the theorem, but are not isomorphic. Let A be the domain of $\mathfrak{A}$ and let B be the domain of $\mathfrak{B}$. Let n be the size of A. Since the fact that A has n elements is expressible by a first-order sentence, B must also have n elements.

Because we assumed that $\mathfrak{A}$ and $\mathfrak{B}$ are not isomorphic, none of the functions $f : A \longrightarrow B$ is an isomorphism. There are only finitely many such functions: f_1, $f_2, f_3, \ldots, f_m$.[1] Also let $a_1, a_2, \ldots, a_n$ be a list of all elements of A.

Because none of the functions f_i, for $1 \leq i \leq n$ is an isomorphism, for each such i there is a formula $\varphi_i(x_1, x_2, \ldots, x_n)$ such that

$$\varphi_i(a_1, a_2, \ldots, a_n) \text{ holds in } \mathfrak{A},$$

but

$$\varphi_i(f(a_1), f(a_2), \ldots, f(a_n)) \text{ does not hold in } \mathfrak{B}.$$

Now, consider the following sentence φ

$$\exists x_1 \exists x_2 \ldots \exists x_n\, [\varphi_1(x_1, x_2, \ldots, x_n) \wedge \varphi_2(x_1, x_2, \ldots, x_n) \wedge \ldots \wedge \varphi_m(x_1, x_2, \ldots, x_n)].$$

The formulas $\varphi_i(x_1, x_2, \ldots, x_n)$ were chosen so that all the conjuncts in φ hold when interpreted by $a_1, a_2, \ldots, a_n$; hence, φ holds in $\mathfrak{A}$. It remains to show that φ does not hold in $\mathfrak{B}$. If it did, it would have been witnessed by a sequence of elements $b_1, b_2, \ldots, b_n$, but then the function $f : A \longrightarrow B$ defined by $f(a_i) = b_i$, for all i such that $1 \leq i \leq n$ is one of the functions $f_j : A \longrightarrow B$ that we have considered, and it follows that $\varphi_j(b_1, b_2, \ldots, b_n)$ does not hold in $\mathfrak{B}$, contradicting φ.

[1] $m = n^n$, but is not important for the argument.

A.4 Existence of Elementary Extensions

In this section we give a proof of the following theorem that played a big role in Chap. 11.

Theorem A.2 *Every structure with an infinite domain has a proper elementary extension.*

All details of the proof are spelled out in the following four paragraphs. It is a formal mathematical argument, but it is simple. The result is very powerful. In Chap. 11, we saw some of its consequences, and it is used in the next section in a proof of an important theorem in combinatorics. The power of the theorem is not in the simple argument below, it comes from the compactness theorem (Fig. A.5).

Let $\mathfrak{A}$ be a structure with an infinite domain A, and let $\overline{\mathfrak{A}}$ be its expansion obtained by adding names for all elements of the domain. We will add to the language one more constant that is not among the constants naming the elements of A. Let us call it c. We will consider a theory T in this expanded language. T consists of the complete theory of $\overline{\mathfrak{A}}$, $\mathrm{Th}(\overline{\mathfrak{A}})$, and the infinite set of sentences of the form $\neg(c = a)$, one for each element a of A. Notice that $\mathrm{Th}(\overline{\mathfrak{A}})$ includes the types of all finite sequences of the elements of A.

It is easy to see that every finite fragment of T has a model. Indeed, if T' is a finite set of sentences from T, then T' consists of a finite number of sentences $\varphi_1, \varphi_2, \ldots, \varphi_m$ that are true in $\overline{\mathfrak{A}}$, and a finite number of sentences of the form $\neg(c = a_1)$, $\neg(c = a_2), \ldots, \neg(c = a_n)$, for some $a_1, a_2, \ldots, a_n$ in the domain. We will show that T' has a model.

Since A is infinite, there is an element in A that is not among $a_1, a_2, \ldots, a_n$. Let us take such an element and interpret the constant c by it. Then all sentences in T' are true $(\overline{\mathfrak{A}}, c)$. We use here the fact that none of the sentences $\varphi_1, \varphi_2, \ldots, \varphi_m$ mentions c.

By the compactness theorem, T has a model. Let us call it $\mathfrak{B}$. All constants naming the elements of A, are interpreted as elements of the domain of $\mathfrak{B}$, and those interpretations are related to each other in $\mathfrak{B}$ exactly as they were in $\mathfrak{A}$, because all the information about those relations is included in T. Hence, $\mathfrak{B}$ contains an isomorphic copy of $\mathfrak{A}$, and in this sense it is an extension of $\mathfrak{A}$. Moreover, since T

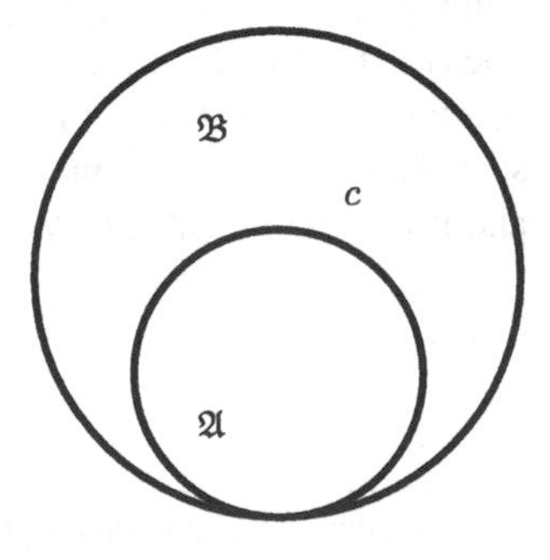

Fig. A.5 $\mathfrak{B}$ is a proper elementary extension of $\mathfrak{A}$ with a new element c

contains types of all finite sequences from A, $\mathfrak{B}$ is an elementary extension of $\mathfrak{A}$. Finally, because the constant c must be interpreted as an element outside the domain of $\mathfrak{A}$, $\mathfrak{B}$ is a proper extension. Its domain has at least one new element.

A.5 Ramsey's Theorem

If an infinite set is cut into a finite number of pieces, then at least one of those pieces must be infinite. How do we know that? This is a basic intuition concerning infinity. We cannot build an infinite set by putting together a finite number of finite sets. If each of the pieces is finite, then so is their union. This section is devoted to a proof of a similar, but harder to prove result. We need some definitions first.

For a set X, $[X]^2$ is the set of all unordered pairs of elements of X. For example, if $X = \{a, b\}$, then $[X]^2$ consists of only one pair $\{a, b\}$.

For a subset A of $[X]^2$, a subset H of X is *A-homogeneous* if either for all distinct a, and b in H, $\{a, b\}$ is in A, or for all distinct a and b in H, $\{a, b\}$ is not in A.

Theorem A.3 (Infinite Ramsey's Theorem for pairs) *Let X be a countably infinite set. For every subset A of $[X]^2$, there exists an infinite subset H of X that is A-homogeneous.*

Ramsey's theorem can be also formulated in term of graphs. Think of the infinite graph whose set of vertices is X, such that there is an edge between any two vertices. Suppose that each edge $\{a, b\}$ is colored either red or blue. Ramsey's theorem says that no matter how the coloring is done, there always will be an infinite set of vertices H such that either all edges $\{a, b\}$, with a and b in H are red, or all such edges are blue.

In another interpretation, think of A as the set of pairs of elements having a certain property. Ramsey's theorem says that for every property of pairs of elements, every infinite set has an infinite subset such that either all pairs from that subset have that property, or none of them have. We can identify each edge with the unordered pair of vertices it connects. This interpretation justifies the inclusion of the theorem in this book, as in this form the theorem is a stepping stone toward constructions of elementary extensions with non-trivial symmetries. The reason to include the proof below is that it illustrates the use of elementary extensions.

The usual proof of Ramsey's theorem is not hard, but is not elementary either. The proof given below is a little simpler than the one that is usually given in textbooks. The simplification is due to the use of Theorem A.2 from the previous section.

A set X is countably infinite if its elements can be enumerated using natural numbers. Thus, to prove the theorem we can assume that X is the set of natural numbers $\mathbb{N}$.[2] The theorem is about sets of unordered pairs of natural numbers, but it will be convenient to represent them as the ordered pairs (m, n), where m is less than n.[3] With these conventions in mind we can now begin the proof .

Let a subset A of $[\mathbb{N}]^2$ be given. We will consider the structure $\mathfrak{N} = (\mathbb{N}, <, A)$ and its proper elementary extension $\mathfrak{N}^* = (\mathbb{N}^*, <^*, A^*)$. Notice that A and A^* are binary relations. Let c be an element of $\mathbb{N}^*$ that is not in $\mathbb{N}$. Because the extension is elementary, for every natural number n, $n <^* c$.[4]

Now we will define a sequence of natural numbers. Let $x_0 = 0$, and suppose that the sequence $x_0, x_1, \ldots, x_n$ has been defined so that $x_0 < x_1 < \cdots < x_n$, and for all $i < j \leq n$

$$(x_i, x_j) \in A \Longleftrightarrow (x_i, c) \in A^*$$

Our task now is to define x_{n+1}. The formula below says that there is an element v of $\mathbb{N}^*$ that behaves with respect $x_0, x_1, \ldots, x_n$ exactly as c does. It is obviously true in $\mathfrak{N}^*$, since c is such an element. To make the statement a bit shorter, we will use symbols ϵ_i, for each $i \leq n$, with the interpretation: if $(x_i, c) \in A^*$, then $\epsilon_i = \in$, otherwise $\epsilon_i = \notin$.

$$\exists v[x_n <^* v \wedge (x_0, v)\epsilon_0 A^* \wedge (x_1, v)\epsilon_1 A^* \cdots \wedge (x_n, v)\epsilon_n A^*]. \qquad (*)$$

Now comes the crucial part of the argument. Since $\mathfrak{N}^*$ is an elementary extension of $\mathfrak{N}$, there must be an element in $\mathbb{N}$ that has the property expressed by $(*)$ when A^* and $<^*$ are replaced with A and $<$, respectively. We define x_{n+1} to be a least such element.

Now imagine that the whole infinite sequence of the $x_0, x_1, x_2, \ldots$ has been constructed. Let $B = \{x_n : n \in \mathbb{N}\}$. The construction guarantees that for all x and y in B, if $x < y$, then (x, y) is in A if and only if $(x, c) \in A^*$. Let $H = \{x \in B : (x, c) \in A^*\}$ if this set is infinite, otherwise let $H = \{x \in B : (x, c) \notin A^*\}$. Then, H has the required property: either for all x in H the pair (x, c) is in A^*, and then for all x and y in H, such that $x < y$, (x, y) is in A; or for all x in H, (x, c) is not in A^*, hence for all x and y in H, such that $x < y$, (x, y) is not in A. This finishes the proof.

---

[2] If $X = \{x_0, x_1, x_2 \ldots\}$ is an enumeration of X, then Ramsey's Theorem for X is a straightforward consequence of Ramsey's Theorem for the set of indices $\{0, 1, 2 \ldots\}$.

[3] Since $\{m, n\} = \{n, m\}$ we can identify each unordered pair with the ordered pair in which $m < n$.

[4] An argument showing this is given in Sect. 12.1.

A.5.1 A Stronger Version of Ramsey's Theorem

The use of an elementary extension and the "infinite" element c in it made the proof of Ramsey's theorem a bit more streamlined than the usual combinatorial proof, but it comes at a price. The proof shows that there must exist a set H with the required properties, but it not telling us much more about this set. That is because we do not have sufficient information about A^* and the "infinte" element c that are used to define the homogeneous set H. The definition of H given in the proof is not a first-order definition in $(\mathbb{N}, <, A)$. It is a set-theoretic definition that refers to the inductively defined infinite sequence $x_0, x_1, x_2, \ldots$. The proof tells us that an infinite homogeneous set exists for any set of pairs A, but we don't know how complex H is in comparison to A. However, there is another similar proof that provides more information.

In Theorem A.3, let us assume in addition that the set A is first-order definable in $(\mathbb{N}, +, \cdot)$. For such an A, we can show that there is a homogeneous set H that is also first-order definable in $(\mathbb{N}, +, \cdot)$. To prove that, instead of an elementary extension $(\mathbb{N}^*, <^*, A^*)$, we can use an elementary extension of the richer structure $(\mathbb{N}^*, +^*, \cdot^*)$. It is not necessary to include A, and $<$ explicitly, since $<$ definable and we assume that A is definable as well. To reach the stronger conclusion, we can now use a stronger version of Theorem A.2. The arithmetic structure $(\mathbb{N}, +, \cdot)$ not only has a proper elementary extension, as any other structure with infinite domain has. It also has a *conservative* elementary extension, i.e. an extension $(\mathbb{N}^*, +^*, \cdot^*)$ such that for each set X^* that is parametrically definable in $(\mathbb{N}^*, +^*, \cdot^*)$, the intersection of X^* with $\mathbb{N}$ is definable in $(\mathbb{N}, +, \cdot)$.[5] If we use such a conservative extension, then one can show[6] that the set H, that we defied in the proof is extended to a set H^* that is defined in $(\mathbb{N}^*, +^*, \cdot^*)$ using the parameter c, so that H is the intersection of H^* with $\mathbb{N}$. It follows that H is definable in $(\mathbb{N}, +, \cdot)$.

[5] In $(\mathbb{N}, +, \cdot)$ all natural numbers are definable, hence every parametrically definable set is also definable without parameters.

[6] This part of the argument is routine, but a bit technical so we skip it.

Appendix B
Hilbert's Program

The general study of mathematical structures by means of formal logic is relatively new. The applications of logic that I described in his book were not the original motivation behind the developments in mathematical logic that led to first-order logic and related formal systems, but they turned out to be a very useful byproduct. Mathematicians study numbers and geometric figures, and come up with patterns and regularities. They deal with *proofs*, irrefutable arguments showing that such and such fact is true. What are those proofs based on? They are based on a *theory* that starts with basic, undeniable observations concerning numbers and geometric figures and then proceeds piling up results derived from previously established ones by logical deductions. That is a description of the kind of mathematics we learn about at school. Some mathematics nowadays is still like that, but most of it is not. While still rooted in classical numerical, geometric, and algebraic problems, modern mathematics, has evolved into an extremely diverse body of knowledge. The current official classification identifies 98 general mathematical subjects, each of which splits into a large number of more specialized subcategories. The names of the subjects will say very little to nonspecialists, and it is becoming harder and harder to have a clear picture of what all this is about. However, there is one distinctive feature that is present explicitly in most subjects, and that is an open and unrestricted use of arguments involving infinity. How does one *use* infinity? If you have studied calculus, you have seen examples of uses of infinity on almost every page the textbook. In fact, the whole idea of calculus is built on reasoning about infinite processes. But we do not need to refer to calculus to see the presence of infinity in mathematics.

Since $\sqrt{2}$ is not rational, it cannot be represented as a finite decimal, and, as shown in Sect. 5.5, it cannot be represented as an infinite repeating decimal. It is represented by an infinite non-repeating sequence of digits. Once we start thinking about it, all kinds of questions arise. In what sense do we say that $\sqrt{2}$ *is* an infinite sequence? How can we know anything about that entire sequence? How is one infinite sequence added to another sequence; how are such sequences multiplied?

R. Kossak, *Mathematical Logic*, Springer Graduate Texts in Philosophy 3,
https://doi.org/10.1007/978-3-319-97298-5

Mathematics provides very good answers, but working with infinite objects requires substantively more than just basic numerical and geometric intuitions. Where do our intuitions concerning infinity come from? All mathematical entities are abstract, but if we only talk about those that are finite (and not too large), we can still rely on our direct insight. We can argue about finite mathematical objects as we would about ordinary "real life" entities, and the results are usually precise and correct. It is not so when it comes to infinity. The history of mathematics is full of episodes showing how cavalier approaches to infinity lead to paradoxes and contradictions.

In the 1870s, Georg Cantor proved that the points filling a square can be put into a one-to-one correspondence with the points of one of its sides. In this sense, the *number* of points of the square is the same as the *number* of points of its side. In Cantor's time, the result was considered counterintuitive, as "clearly" two-dimensional objects (squares) are larger than one-dimensional ones (line segments).

Cantor's result is just a first step into the exploration of the vast world of infinite sets. Infinitistic methods proved useful in establishing mathematical results, but they were also a subject of criticism and concerns about their validity. In response to criticism, Hilbert proposed a program to prove once and for all that infinity, and set theory that deals with it, have a proper place in mathematics. Here is an outline of his program:

1. Define a system based on a formal language in which all mathematical statements can be expressed, and in which proofs of theorems can be carried out according to well-defined, strict rules of proof.
2. Show that the system is *complete*, i.e. all true mathematical statements can be proved in the formalism.
3. Show that the system is *consistent*, i.e. it is not possible to derive a statement and its negation. The proof of consistency should be carried out using finitistic means without appeal to the notion of actual infinity.
4. Show that the system is *conservative*, i.e. if a statement about concrete objects of mathematics, such as natural numbers or geometric figures, has a proof involving infinitistic methods, then it also has an elementary proof in which those methods are not used.
5. Show that the system is *decidable* by finding an algorithm for deciding the truth or falsity of any mathematical statement.

All elements of Hilbert's program had a tremendous impact on the development of the foundations of mathematics. Our main concern in this book was its first, initial stage—designing a system in which all of mathematics can be formalized. It is the part of the program that has been successfully completed in the first half of the twentieth century. At the beginning, a major candidate for the formal system to encompass all mathematics was the theory of types developed by Alfred North Whitehead and Bertrand Russell in *Principia Mathematica*. The elaborate system of Whitehead and Russell was later replaced by a combination of first-order logic, developed earlier by Gottlob Frege, and axiomatic set theory. The story of Hilbert's remaining desiderates is complex and interesting. First-order logic turned out to be complete, as was shown by Kurt Gödel in 1929; but in the following year

Gödel proved his celebrated incompleteness theorems which in effect caused the collapse of Hilbert's program. Gödel showed that for any consistent formal system that is strong enough there are statements about natural numbers that are true, but unprovable in the system. Hilbert's program collapsed, but it inspired a great body of research in foundations of mathematics, including all methods we discussed in this book. Of particular interest is the consistency problem.

If we argue informally, it is not unusual to come up with contradictory statements. A good example is Cantor's theorem stating that there is a one-to-one correspondence between the points of a square and the points of one of its sides. When Cantor proved his result, it was believed to be false, and in fact it was later proved to be false by L.E.J. Brouwer. How is that possible? How can one have proofs of two contradictory statements? The secret is that when nineteenth century mathematicians talked about one-to-one correspondences, they meant not arbitrary correspondences, but the continuous ones. Brouwer proved that there cannot be a continuous one-to-one correspondence between the points of a square and the points of its side, confirming the good intuition of the mathematicians who had thought that to be the case. Cantor's one-to-one correspondence was (and had to be) discontinuous. Contradictions in informal reasoning can arise from imprecise understanding of terms used. Once differences in terminology are clarified, contradictions should disappear. There is no guarantee that this will always happen, but at least this is how it has worked in practice so far.

In formal systems all terms are precisely defined upfront, and so are the rules of arguments (proofs). There is no room for ambiguity. There are many proof systems, but their common feature is that the rules of proof are mechanical. How do we design such systems, and how can we prove that inconsistencies will not be found in them? It was Hilbert's profound insight that formal statements and formal proofs are finite combinatorial objects and that one can argue about them mathematically. If there is no contradiction in a system (which we believe to be the case for the commonly used systems of mathematics today), this itself is an example of a mathematical statement that can be turned into a theorem once a proof is supplied. Hilbert believed that this could be done. His belief was shattered by Gödel's second incompleteness theorem which states that if a system is strong enough its consistency cannot be proved within the system. Formalized mathematics cannot prove its own consistency!

It was only a brief sketch, for a full account I strongly recommend Craig Smoryński's excellent book [32].

While Hilbert's program turned out untenable, its impact can not be overstated. Formalization of mathematics was crucial for the development of computer science and computer technology. We can talk to machines and they talk to us. We have a formal language to communicate. This is a success story, but in this book we have examined a different, and perhaps somewhat unexpected aspect of formalization. One could ask: Is there any point in learning a formal language if one is not interested in formalizing mathematical arguments or in talking to a machine? The answer is "yes," for several reasons. One is that formal methods play an important role in building and analyzing mathematical structures, and if you are interested in structures in general, those are good examples to consider. Secondly, when expressed in a formal language, properties of structures become, as Hilbert wanted, mathematical objects. Each has its own internal structure. They can be classified according to various levels of complexity. They can be manipulated revealing meanings that may not be that transparent in an informal approach. Finally, the properties of elements of structures that are expressible by first-order formulas have a certain *geometric structure*. This geometry brings a kind of geometric thinking that can be applied in situations that do not seem to have much to do with traditionally understood geometry. Such geometrization has been effective in several areas of modern mathematics. For an account of these developments, see David Marker's entry on Logic and Model Theory in [23].

Bibliography

1. Badiou, A. (2010). *Being and event*. London/New York: Continuum.
2. Bagaria, J. Set theory. In E. N. Zalta (Ed.), *The Stanford encyclopedia of philosophy* (Winter 2017 ed.). https://plato.stanford.edu/archives/win2017/entries/set-theory/
3. Baldwin, J. T. (2017). *Model theory and the philosophy of mathematical practice: Formalization without foundationalism*. Cambridge: Cambridge University Press.
4. Button, T., & Walsh, S. (2018). *Philosophy and model theory* (With a historical appendix by Wilfrid Hodges). Oxford: Oxford University Press.
5. Beth, E. W. (1970). *Aspects of modern logic*. Dordecht/Holland: D. Reidel Publishing Company.
6. Borovik, A. (2014). English orthography as a metaphor for everything that goes wrong in mathematics education. *Selected Passages from Correspondence with Friends, 2*(2), 9–16.
7. DeLillo, D. (2016). *Zero K*. New York: Scribner.
8. Hacking, I. (2014) *Why is there philosophy of mathematics at all?* Cambridge: Cambridge University Press.
9. Hallett, M. Zermelo's axiomatization of set theory. In E. N. Zalta (Ed.), *The Stanford encyclopedia of philosophy* (Winter 2016 ed.). https://plato.stanford.edu/archives/win2016/entries/zermelo-set-theory/
10. Hardy, G. H. (1929). Mathematical proof. *Mind. A Quarterly Review of Psychology and Philosophy, 38*(149), 1–25.
11. Hilbert, D. (1926). Uber das Unendliche. *Mathematische Annalen, 95*, 161–190.
12. Dews, P. Review of *Being and Event* by Alain Badiou. *Notre Dame Phiosophical Reviews*. https://ndpr.nd.edu/news/being-and-event/
13. Hodges, W. Tarski's truth definitions. In E. N. Zalta (Ed.), *The Stanford encyclopedia of philosophy* (Fall 2014 ed.). http://plato.stanford.edu/archives/fall2014/entries/tarski-truth/
14. Hodges, W. Model theory. In E. N. Zalta (Ed.), *The Stanford encyclopedia of philosophy* (Fall 2013 ed.). https://plato.stanford.edu/archives/fall2013/entries/model-theory/
15. Hodges, W., & Scanlon, T. First-order model theory. In E. N. Zalta (Ed.), *The Stanford encyclopedia of philosophy* (Spring 2018 ed.). https://plato.stanford.edu/archives/spr2018/entries/modeltheory-fo/
16. Husserl, E. (2003). *Philosophy of arithmetic. Psychological and logical investigations—With supplementary texts from 1887–1901* (Edmund Husserl Collected Works, Vol. X, D. Willard, Trans.). Dordrecht/Boston: Kluwer Academic Publishers.
17. Grzegorczyk, A. (2013). *Outline of mathematical logic: Fundamental results and notions explained with all details*. New York: Springer.

R. Kossak, *Mathematical Logic*, Springer Graduate Texts in Philosophy 3,
https://doi.org/10.1007/978-3-319-97298-5

18. Kline, M. (1972). *Mathematical thought from ancient to modern times*. New York: Oxford University Press.
19. Knight, J. F., Pillay, A., & Steinhorn, C. (1986). Definable sets in ordered structures. II. *Transactions of the American Mathematical Society, 295*(2), 593–605.
20. Mancosu, P. (2012). *The adventure of reason. Interplay between philosophy of mathematics and mathematical logic: 1900–1940*. Oxford: Oxford University Press.
21. Mancosu, P. (2016). *Abstraction and infinity*. Oxford: Oxford University Press.
22. Marker, D. (2002). *Model theory: An introduction* (Graduate texts in mathematics, Vol. 217). New York: Springer.
23. Marker, D. (2008). Logic and model theory. In T. Gowers, J. Barrow-Green, & I. Leader (Eds.), *Princeton companion to mathematics*. Princeton: Princeton University Press.
24. Mostowski, A. (1969). *Constructible sets with applications*. Amsterdam: North Holland Publishing Company/Warszawa: Państwowe Wydawnictwo Naukowe.
25. Manin, Y. (2009). *A course in mathematical logic for mathematicians* (Graduate texts in mathematics, Vol. 53, 2nd ed., with Collboration by Boris Zilber). New York: Springer.
26. Mostowski, A. (1967). *Thirty years of foundational studies: Lectures on the development of mathematical logic and the study of the foundations of mathematics in 1930–1964* (Acta Philosophica Fennica, Vol. 17). Helsinki: Akateeminen Kirjakauppa.
27. Mostowski, A. (1979). *Foundational studies: Selected works*. North-Holland: Elsevier.
28. Peterzil, Y., & Starchenko S. (1996) Geometry, calculus and Zil'ber's conjecture. *Bulletin of Symbolic Logic, 2*(1), 72–83.
29. Reid, C. (1996). *Julia, a life in mathematics* (With Contributions by Lisl Gaal, Martin Davis and Yuri Matijasevich). Washington, DC: Mathematical Association of America.
30. Reid, C. (1996). *Hilbert*. New York: Springer.
31. Singh, S. (1997). *Fermat's last theorem*. London: Fourth Estate.
32. Smoryński, C. (2012) *Adventures in formalism*. London: College Publications.
33. Stillwell, J. (2018). *Reverse mathematics. Proofs from the inside out*. Princeton: Princeton University Press.
34. Tarski, A. (1933). The concept of truth in the languages of the deductive sciences (Polish). *Prace Towarzystwa Naukowego Warszawskiego, Wydział III Nauk Matematyczno-Fizycznych, 34*, Warsaw; expanded English translation in Tarski, A. (1983). *Logic, semantics, metamathematics, papers from 1923 to 1938*. Edited by John Corcoran. Indianapolis: Hackett Publishing Company.
35. Tarski, A. (1948). *A decision method for elementary algebra and geometry*. Santa Monica: RAND Corporation.
36. Van den Dries, L. (1998). *Tame topology and o-minimal structures*. New York: Cambridge University Press.
37. Wagon, S. (1985). *The Banach-Tarski paradox*. Cambridge: Cambridge University Press.
38. Wittgenstein, L. (1976). *Wittgenstein's lectures on the foundations of mathematics, Cambridge, 1939* (Edited by Cora Diamond). Chicago: The University of Chicago Press.
39. Zalamea, F. (2012). *Synthetic philosophy of contemporary mathematics*. Falmouth: Urbanomic/New York: Sequence Press.

Index

R. Kossak, *Mathematical Logic*, Springer Graduate Texts in Philosophy 3,
https://doi.org/10.1007/978-3-319-97298-5

Printed in the United States
By Bookmasters

Printed in the United States
By Bookmasters